Ground Station Design and Analysis for LEO Satellites

Ground Station Design and Analysis for LEO Satellites

Analytical, Experimental and Simulation Approach

Shkelzen Cakaj

IEEE PRESS

WILEY

For general information on our other products and services or for technical support, please contact our Customer Care Department within the United States at (800) 762-2974, outside the United States at (317) 572-3993 or fax (317) 572-4002.

Wiley also publishes its books in a variety of electronic formats. Some content that appears in print may not be available in electronic formats. For more information about Wiley products, visit our web site at www.wiley.com.

Library of Congress Cataloging-in-Publication Data applied for
Hardback ISBN: 9781119899259

Cover Design: Wiley
Cover Image: © Rob Atkins/Getty Images

Set in 9.5/12.5pt STIXTwoText by Straive, Pondicherry, India

*To my wife Naime and our children
Gresa, Vesa and Genti*

Shkelzen

Contents

Preface

I belong among the experts who have combined their professional career in the telecommunications industry with academic scientific research. I have been involved for 40 years in the development and business processes in Kosovo's telecommunication system, while for the last 20 years I was also engaged in academic scientific activities at universities in Republic of Kosovo and Republic of Albania. My entire scientific research interest was dedicated to the performance of ground stations for LEO (low Earth orbiting) satellites, with an outcome of 40 scientific published papers worldwide, exclusively related to LEO ground stations.

Throughout my career, I have been involved in several technological telecommunication infrastructure generations, each one providing significant improvements, enhancing the services toward a better social future. Thus, to comprehend the technological effect into our lives, one could simply compare the lifestyle of our grandparents with that of our nieces and nephews, which covers the period of approximately 100 years. The difference is unimaginable. From my perspective, this tremendous difference, including the current focus on worldwide communications, stems from the two crucial technological achievements from the past century – the satellite systems and Internet. The interoperation of both, enabled impressive long distance multimedia communication services, virtual meetings and the long-distance social interoperability.

Recently, significant international efforts are oriented toward the framework definition for a global satellite communication system which would be fully integrated within the existing terrestrial system. The most convenient structures for such a development are the LEO satellites. The latter operate closer to Earth compared to the other orbits, providing significantly lower latency, which is crucial for reliable and safe communications. The LEO satellites combined with the ground stations as a part of the satellite-terrestrial integrated network, through their interoperability, are the future of communication infrastructure, intending to offer the global Earth coverage with broadband multimedia services, from South Africa to Tibet, from Alaska to New Zealand. But such an approach involving many satellites will result in new scientific challenges, not only in the integrated satellite-terrestrial networks, but also pertaining to the sky transformation itself.

Except for communication purposes, LEO satellites are seriously and effectively applied for scientific missions as well. Potential applications are vast, including but not limited to the remote sensing of oceans, analyses of Earth's climate changes, mapping, Earth's imagery with high resolution, navigation, management of Earth's resources, astronomy, military, agriculture, and even humanitarian efforts for search and rescue services.

Therefore, it is to be expected that such missions will be further developed in the future, especially in the fields where similar experiments cannot be done by means on the Earth. Thus, ground stations must be established to communicate with satellites to further these missions. Their performance is crucial for reliable communication; thus, they must be carefully and multidimensionally analyzed, which is the subject of this book. The same approach should be applied to the access points on ground dedicated for internet access.

My passion for LEO satellites began when I joined the LEO satellite MOST (Microvariability and Oscillations of Stars) ground station at the Vienna Technical University in Austria. In September 2003, in the capacity of a scientific guest, I joined a working group at the Institute for Communication and Radiofrequency Engineering of Technical University in Vienna. The group was tasked to work on the implementation of a satellite ground station in Vienna, dedicated for communication with a Canadian space observation LEO microsatellite – MOST.

For more than three years starting in 2005, I worked with the Department for Radio-Communication and Microwave Engineering at the Zagreb University in Croatia, focusing on my PhD thesis and analyzing the satellite ground station performances in the urban areas. These analyses, titled "Rigorous Analysis on Performance of LEO Satellite Ground Station in Urban Environment," are published in the *International Journal of Satellite Communications and Networking*, and form the core scientific contribution of my PhD thesis.

In 2009, I was supported by the Fulbright program for my postdoctoral research. The research area treated the simulation and implementation protocols for local user terminals dedicated for search and rescue services supported by satellites. I spent three months working at the United States National Oceanic and Atmospheric Administration (NOAA), analyzing the performance of the local user terminals and applying simulations for hypothetical distress events resulting into conclusions about the performance of local user terminals under different circumstances.

My overall scientific engagement over two decades related to the satellite ground stations performance dedicated for LEO satellites, is fourfold: (i) atmospheric impairments, (ii) coverage area from LEO satellites, (iii) ground station's ideal and designed horizon plane, and (iv) communication duration optimization between LEO satellites and appropriate ground stations. All of these aspects interlinked together are also treated, for the overall performance evaluation of the ground station, usually expressed through ground stations' Figure of Merit.

In March 2021, I contracted COVID-19 and became severely ill. I fought to survive for several weeks. In moments of subtle hope, I promised myself that if I got out of bed, I would condense my scientific work in the form of a single book. COVID touched and woke up my motivational instinct for publishing this book!

This book reflects the consolidated research of 20 years, including mathematical analyses, experiments, and simulations that could serve as a guide for the ground station performance evaluation at any of the worldwide LEO stations. My approach focuses on four key components: idea, methodology, results, and conclusions stemmed from my research, as an innovative tutorial and an advanced level guide to ground stations analysis and design. In my view, this makes the difference!

Usually, in the literature, the satellite ground stations are treated as a chapter within a book for satellite communications or more generally for satellite systems. My intention through this book is to bring to the readers a deeper insight, not only about the satellite ground station organization but also about the performance of each separate block and the entire ground station performance evaluation toward the safe and reliable functionality, considering technical characteristics of devices and the environmental circumstances.

This book is organized into 10 chapters and ends with a few short, final remarks. Being aware that there are readers who, for different reasons, might not be able to read the entire book, I have been careful that each chapter is organized and compacted in such a manner that it can also be read as single one, designed to provide the necessary and expected information for the respective readers.

Chapter 1 covers the most general concepts of the satellite ground station organization, providing an overview of the single and double antenna system configuration. Ground station subsystems and respective components are described, followed by the Figure of Merit interpretation, including both ground station equipment parameters and external environmental factors impact. For purposes of illustration, the chapter provides a brief description of the Canadian satellite MOST (Microvariability and Oscillation of STars) and the respective ground station implemented in Vienna, which is further applied for the interpretation purposes throughout the entire book.

Chapter 2 focuses on rain attenuation since it is considered as the most impactful atmospheric factor on the radio waves propagation. Behind the general aspect of the rain attenuation, based on the rain attenuation path geometry, the modeling of rain attenuation is provided. The modeling approach is applied for the rain attenuation calculation for different European cities randomly chosen. The appropriate data provide hints about the atmospheric impairments attenuation to be applied due to the link budget related calculations anywhere.

The downlink performance is expressed through the downlink budget. The Figure of Merit is the main downlink performance indicator, depending on the implemented equipment characteristics and environmental factors, represented by system noise temperature. System noise temperature consists of composite noise temperature and antenna noise temperature. In Chapter 3, two scenarios of the downlink budget are compared – for the single antenna and double antenna configurations. This chapter ends with the experimental measurement of the Figure of Merit, based on the Sun Flux Density, experimentally confirming the mathematical calculations previously applied.

Chapters 4 and 5, respectively, consider the ground station horizon plane and the coverage area of the LEO satellites, issues that are closely interrelated. The coverage area is the fraction of Earth's surface covered by the LEO satellite and the horizon plane determines the communication zone between the satellite and the appropriate ground station. Three types of ground station horizon planes are analyzed in details, the ideal, practical, and designed one. The mathematical and geometrical correlation between the ideal and the designed horizon plane are established as well. The time efficiency factor is implemented in order to quantify the difference in communication duration under ideal and designed horizon plane. Further, the individual and global coverage from the LEO satellites is clarified. This is accompanied with specifically calculated coverage for the LEO satellites, proving their low coverage area. Chapter 5 also includes a geometrical confirmation of the handover process due to the global coverage by the LEO constellation.

In Chapter 6, the Sun's synchronized orbits are analyzed, as they are very useful for the photo imagery missions. The inclination window for LEO altitudes to be Sun synchronized is calculated. Further, under that inclination window, the perigee deviation for the Sun synchronized orbits is interpreted. Chapter 7 presents the launching satellite procedure and discusses coplanar Hohmann transfer analysis under the different LEO altitudes. Chapter 7 ends with a detailed procedure of geostationary altitude attainment form the Kourou launching site.

The LEO satellites applied for the search and rescue missions are considered in Chapter 8. Simulation of local user terminals dedicated for search and rescue services is provided, assuming the hypothetical local user terminal and four randomly supposed distress locations. The existing LEO-SAR and the future to be finalized MEOSAR constellation are compared. Toward the end of the book, the interference issue is considered throughout Chapter 9. The intermodulation products are analyzed and the modeling for in advance interference identification is given. The adjacent interference is analyzed through the typical example of the interference avoidance for the case of the NOAA constellation. Modulation index application for the interference identification accompanied with practical records is provided in this chapter as well. Finally, Chapter 10 outlines the need for further scientific research. The first one relates to free space loss compensation through dynamic bandwidth tunability, while the second relates to the wideness horizon plane parameter to be applied in the future for the analysis of the LEO ground station performance.

The book ends with "Closing Remarks," in which I emphasize the three most important events (in my opinion) in satellite systems in the last 10 years. The first one related to developments of April 2014, when the first stage of the Falcon 9 rocket had made a controlled power landing on the surface of the Atlantic Ocean. The second relates to the Orbit Fab, a San Francisco-based space-industry startup, that has developed end-to-end refueling service using its Rapidly Attachable Fluid Transfer Interface (RAFTI). Finally, the third one is related to the launching of satellites from the International Space Station (ISS). These achievements, beside economic advantages, will also support the longer satellite life in space, for the better and longer coexistence with terrestrial systems.

I retired in the middle of March 2022, but I am still engaged in the education system in Republic of Kosovo and Republic of Albania, teaching new generations about the principles and the future of the satellite communications. In the future, I will still keep my friendship with LEO satellites.

Acknowledgments

I would like to deeply thank Professor Arpad L. Scholtz and Dr. Werner Keim for their support and excellent cooperation during my time spent at the Institute for Communications and Radiofrequency Engineering of Technical University in Vienna, Austria, to study satellite ground station performance in Vienna for communication with the MOST satellite. My gratitude also goes to Professor Kresimir Malaric from the faculty of Electrical and Computing Engineering of Zagreb University, Croatia, for his excellent guidance as I worked on my doctoral thesis. For my postdoc research, as Fulbright alumnus, at NOAA-SARSAT-NSOF in Suitland, Maryland, US, related to the application of LEO satellites in search and rescue services, I acknowledge Micky Fitzmaurice, for support and open cooperation. I recognize my colleagues, Professor Bexhet Kamo and Professor Elson Agastra, from the faculty of Information Technology of Polytechnic University of Tirana, Albania, for their careful reading of the manuscript and valuable suggestions. Finally, I would like to acknowledge senior editor Sandra Grayson for her support from my first idea for this book and also Kimberly Monroe-Hill and Becky Cowan for carefully managing the editing and technical support, respectively.

1

LEO Satellite Ground Station Design Concepts

1.1 An Overview of LEO Satellites

Satellites are an important part of telecommunication infrastructure worldwide, carrying large amounts of multimedia traffic. Since their inception around 60 years ago, communication satellites have been a major element in worldwide communication infrastructure and networking. More than 40 countries own satellites for communication, commercial, science, and even humanitarian purposes. But only few of them have building and launching capabilities.

The basic resources available for satellite communications are orbits and radio frequency (RF) spectrum. The orbit is the path in space followed by the satellite; the frequency allocations are subject to international agreements managed and controlled by international bodies.

Different types of orbits are possible, each suitable for a specific application or mission. Generally, satellites follow an elliptical orbit with a determined eccentricity laid on the orbital plane defined by space orbital parameters (Maral and Bousquet 2005; Maini and Agrawal 2011). Thus, the space orbital parameters, known as Kepler's elements (usually given as two-lines elements), determine the position of the satellite in space (space slot). Orbits with zero eccentricity are known as circular orbits. The circularity of the orbit simplifies the analysis, compared to the elliptical one. The movement of the satellite within its circular orbit is represented by altitude, radius velocity, and orbital time.

Satellites' circular orbits are categorized as geosynchronous Earth orbits (GEO), medium Earth orbits (MEO), and low Earth orbits (LEO). The main difference among them is in the altitude above the Earth's surface, which further impacts the velocity and the orbital period of the satellite in the appropriate orbit (Maral and Bousquet 2005; Maini and Agrawal 2011). Only the circular orbits are of the further concern through this book, more exactly, the LEO satellites and the appropriate ground stations.

Communication between the satellite and a ground station is established when the satellite is consolidated in its own orbit and it is visible from the ground station. The link that transmits radio waves from the ground station to the satellite is called uplink, and from satellite toward the ground station is downlink.

The orbits of altitudes ranging from 300 km up to around 1400 km above the Earth's surface are defined as LEO, and the satellites consolidated to these orbits are known as the LEO satellites. The lower altitude range is limited by the Earth's atmosphere – more accurately, by the level above the Earth's atmosphere where there is almost no air, so the satellite's speed reduction and drag down is avoided. The inner Van Allen belt limits the higher altitude range (Van Allen radiation belt 2020). The Van Allen belt is known as a space radiation zone and has undesired effects on satellites'

Ground Station Design and Analysis for LEO Satellites: Analytical, Experimental and Simulation Approach,
First Edition. Shkelzen Cakaj.
© 2023 The Institute of Electrical and Electronics Engineers, Inc. Published 2023 by John Wiley & Sons, Inc.

payload and platform (electronic components and solar cells can be damaged by this radiation); thus, the belt should not be used for the accommodation of LEO satellites.

LEO satellites move at around 7.2–7.5 km/s velocity relative to a fixed point on the Earth (ground station). Satellites' orbital period is in the range of 90–110 minutes. The communication duration between the satellite and the ground station takes 5–15 minutes over 6–8 times during the day (Cakaj and Malaric 2007a), all these dependent on orbital altitude. The characteristics of LEOs are the shortest distance from the Earth compared with other orbits and consequently less time delay. These characteristics make them very attractive for communications but also for other applications (Cakaj 2021).

Thus, in addition to communications, LEO satellites are also applied for scientific and research purposes, more specifically under circumstances where no on-ground means are appropriate. Dynamics on climate changes, remote sensing applications for oceans, different astronomic observations, ions density records in the ionosphere, and very specific humanitarian applications related to search and rescue services are some of activities carried out by LEO satellites, activities that are too difficult or impossible to be implemented on Earth. For these activities within satellite structures, the instruments or devices (telescope, cameras, probes, sensors, etc.) for the appropriate application or mission are installed (Zee and Stibrany 2002; Cakaj et al. 2010a). Usually, LEO satellites dedicated for scientific purposes or remote sensing applications are accommodated in specifically designed orbits, known as the Sun synchronized orbit. The Sun synchronization feature enables a treated area on the ground from the satellite to be observed under similar illumination conditions due to different satellite passes (Cakaj et al. 2009).

These satellites provide opportunities for investigations for which alternative techniques are either difficult or impossible to apply. Thus, it may be expected that such missions will be further developed soon, especially in fields where similar experiments by purely Earth-based means are impracticable. Ground stations (access points) must be established to communicate with such satellites, and the quality of communication depends on the performance of the satellite ground station, in addition to that of the satellite.

Communications-integrated satellite-terrestrial networks used for global broadband services have gained a high degree of interest from scientists and industries worldwide. The most convenient structures for such use are LEO satellites, since they fly closer to the Earth compared to the other orbits, and consequently provide significantly lower latency, which is essential for reliable and safe communications. Among these efforts is the *Starlink* satellites constellation, developed and partly deployed by the US company SpaceX. The constellation is planned to be organized in three spatial shells, each made up of several hundreds of small-dimensioned and lightweight LEO satellites specially designed to provide broadband services, intending to offer global Earth coverage through their interoperability, combined with the ground stations as a part of the satellite-terrestrial integrated network. On October 24, 2020, 893 satellites were situated in orbit of altitudes of 550 km under different inclinations, determining the first *Starlink* orbital shell (Cakaj 2021).

This would suggest that in the near future, worldwide broadband services provided by integrated satellite-terrestrial communication networks will be a part of daily communication activities, demands for which will rapidly increase, so operators should carefully manage operation and distribution of real-time services toward maximizing the downlink data throughput related to the broadband requirements without significantly affecting the mission cost (Botta and Pescape 2013; Garner et al. 2009). Therefore, future satellite payloads and platforms must become more flexible, lightweight, and smaller, easier to be launched, and reconfigurable related to the EIRP and coverage, to provide large capacity at the lowest cost, toward the main goal of the worldwide coverage with broadband services and other scientific missions, as well.

According to the worldwide coverage missions, their network architecture in space could be categorized into single-layer (one-shell) networks and multilayer networks. A single-layer network provides intercommunication between only satellites of the same altitude, whereas multilayer networks enable communications between satellites in different shells. Multilayer networking is more complex but is advocated for its flexibility in providing more sustainable global coverage, seamless handovers, and reliable communications.

LEO satellites used at the end of the past century were known as microsatellites because of their light weight and small dimensions. Later, nanosatellites were developed as more convenient structure for launching process, since less energy is required to launch such satellites into the LEO space slot. But recently, it has been possible to launch nanosatellites from the International Space Station (ISS) (List of spacecrafts deployed from the International Space Station 2020). Related to the launching process, LEOs play an additional role as the first space shell for the satellites toward geosynchronous (geostationary) orbits, due to the three-step transfer process (known as Hohmann transfer) (Cakaj et al. 2015a).

LEO satellites and appropriate ground platforms (access points) now represent a very useful system, not only for the main mission as communication is but also for research scientific missions. Through LEO satellites and appropriate platforms, anywhere on the globe can be provided data about the water dirtiness of the river Amazon, about new exoplanets, natural disasters, air or marine disasters, how the wheat is growing in South Africa, how many refugees are crossing the borders, ice melting, and increasing seawater level, for example.

Related to the last item, the satellite *Sentinel-6 Michael Freilich*, launched on November 21, 2020, from Vandenberg launching site in California and consolidated into the LEO orbit of altitude of 1336 km under 66^0 inclination, will measure the sea level around the globe for the next five years. The mission is collaboration between NASA and the European Space Agency (see Figure 1.1) (Sentinel-6 Michael Freilich 2021).

Finally, as the nineteenth century was deeply marked with the steam machines, this century will be marked by LEO satellites, hopefully for the better life on Earth! These tools provide opportunities not only for communications but also for scientific purposes, including Earth and space observation. LEO satellites serving as "eyes" in the sky might also prove useful for world peace!

Figure 1.1 *Sentinel-6 Michael Freilich* spacecraft.

Communication with such missions is enabled through the ground stations; thus, the performance of the ground station is crucial for such missions, and will be elaborated on throughout this book.

1.2 Satellite System Architecture

The scheme of a typical satellite communication system architecture is shown in Figure 1.2 (Maral and Bousquet 2005). It includes a ground segment, space segment, and control segment.

The operational satellite receives the radio waves transmitted by the ground station. This is called *uplink*. The received signals by satellite are processed, translated into another radio frequency, and amplified on-board. In turn, these signals are further transmitted to the receiving ground station. This is called *downlink*. Uplinks and downlinks are based on radio frequency modulated carriers' principles. Carriers are modulated by baseband signals, including analog or digital, conveying information for communication or for other purposes.

The *space segment* contains one or several active and spare satellites organized in a constellation. The satellite is an artificial body orbiting around the Earth as "flying" trans-receiver, either for communication or scientific purposes. Each satellite consists of a *payload* and *platform* (bus). The *payload* consists of the receiving and transmitting antennas and all electronics that support the reception and the transmission of radio carriers. The satellite's payload has two main functions:

To amplify the received carriers for retransmission to the downlink. Large distance between the ground station and the satellite causes the carrier's power at the input of the satellite's receiver to be too low. Thus, power must be amplified to feed the satellite's transmit antenna toward users on ground within its coverage area.

Frequency conversion. Frequency conversion is required to increase isolation between the receiving input and transmitting output (avoiding the re-injection into the receiver). In Figure 1.3, the transparent satellite payload is given, making clear the uplink/downlink isolation.

Transparent payload belongs to a single antenna beam satellite where each transmit and receive antenna generates only one beam. Figure 1.3 shows that carriers are power amplified, and frequency

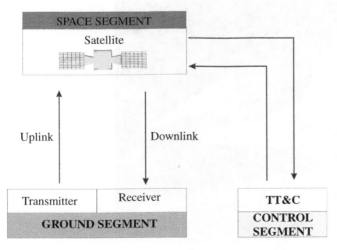

Figure 1.2 Typical satellite communication system architecture.

Figure 1.3 Transparent payload.

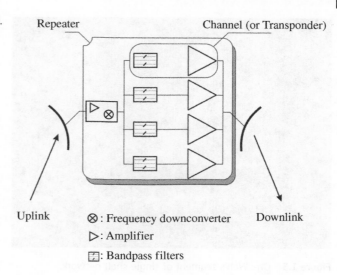

Repeater

Channel (or Transponder)

Uplink

⊗ : Frequency downconverter

▷ : Amplifier

☒ : Bandpass filters

Downlink

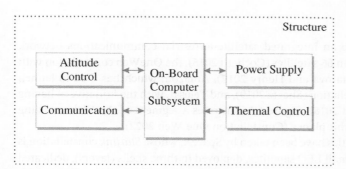

Figure 1.4 Satellite platform block scheme.

down converted. The amplifying chain associated with each sub-band is called *satellite channel* or *transponder*. The bandwidth splitting is achieved using a set of filters. Regenerative payload (multibeam) antennas would have many inputs/outputs as up beams/down beams. Routing of carriers from one up beam to a given down beam implies on-board switching at radio frequency. LEO satellites for scientific purposes usually use single-beam antennas.

The satellite *platform* consists of subsystems that permit the payload to operate. These subsystems are: structure, power supply, temperature control, altitude control, and communication subsystem (see Figure 1.4).

The structure provides the necessary mechanical support. The electrical power supply subsystem provides the necessary DC power. The altitude control subsystem stabilizes the satellite and controls its orbit. The thermal control system maintains the temperature of various subsystems within tolerable limits. All these functions are controlled by on-board computerized subsystem.

Some missions, in principle, can be realized just by a single satellite, but for real-time continuity of services and large or full Earth's coverage, the space segment must be organized as single-layer or multilayer constellation. A single-layer network provides intercommunication between only satellites of the same altitude, whereas multilayer networks enable communications between satellites in different orbital shells. Multilayer networking is more complex, but it is more preferable.

Figure 1.5 One Web's segment of single shell network.

Active satellite projects related to an integrated satellite-terrestrial communications network include the Iridium constellation with 66 satellites (Cochetti 2015), the OneWeb constellation with 648 satellites (De Selding 2015; Pultarova and Henry 2017), Amazon, which has filed to launch 3,236 spacecrafts in its Kuiper constellation (Sheetz 2019), and Telesat, with the initiative of having a 117-spacecraft constellation (Foust 2018). Figure 1.5 illustrates a segment of OneWeb's existing network showing its longitudinal orbit planes (Constellation One Web 2022).

In my view, the most serious activities have been taken by SpaceX, whose *Starlink* constellation is planned to consist of thousands of small LEO satellites, deployed in three shells (layers), dedicated to maximizing broadband internet services toward global Earth coverage, and combined with ground stations (trans-receivers), to be organized as a satellite-terrestrial integrated network for real time worldwide broadband services. The *Starlink* single layer constellation at altitude of 550 km is given in Figure 1.6 (The real benefit of SpaceX-*Starlink,* highspeed internet, 2022).

The *ground station* is the location on the ground equipped with appropriate equipment to be used for communication with the satellite. The function of a ground station is to receive or transmit the information from/to the satellite in the most reliable manner while retaining the desired signal quality at the destination. Scientific missions can be accomplished in principle by only one ground station. The reason behind building more ground stations is to increase the coverage and number of measurements per observed objects or area, and practically increase data download capability. The communication between the satellite and a ground station is established when the satellite is consolidated in its own orbit, and it is visible from the ground station.

The ground segment consists of all the ground stations. These stations are most often connected to the end user equipment by a terrestrial network. Stations are distinguished by their size, which varies according to the volume of traffic to be carried and the type of traffic (voice, video, or data). Ground stations have experienced a tremendous reduction in size. The largest ground stations are equipped usually with antennas of 30 m diameter (Standard A of the INTELSAT network). The smallest ground stations have typically 0.6 m antennas (direct television receiving stations). Some stations both transmit and receive, and some of them are just receive-only (RCVO) stations.

The general organization of a ground station consists of antenna subsystem with associated tracking system, transmitting and receiving equipment, monitoring system, and normally power supply.

Figure 1.6 Starlink satellite single shell constellation at altitude of 550 km.

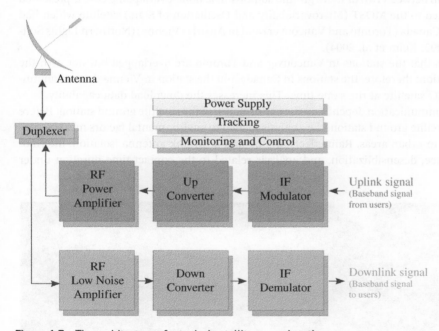

Figure 1.7 The architecture of a typical satellite ground station.

Figure 1.7 shows typical architecture of a ground station for both receiving and transmitting branches. This is a single antenna system where the uplink and downlink separation is achieved by duplexer (Cakaj and Malaric 2007a).

The *control segment* consists of all ground facilities for the control and monitoring of the satellites. This is known as Tracking, Telemetry & Command (TT&C).

1.3 The Satellite Ground Station

Ground stations are vital elements in any satellite communications network. Generally, they serve as an interface for communication between the satellite and different customers. The function of a ground station is to receive information from, or transmit information to, the satellite in the most cost effective and reliable way, while retaining the desired signal quality.

Depending on the applications, the ground stations may have both transmit and receive capabilities or may only be capable of either transmission or reception. Further categorization can be based on the type of services. Usually, the design criteria are different for the Fixed Satellite Service (FSS), the Broadcast Satellite Service (BSS), and Mobile Satellite Service (MSS).

Further concern is related to the ground stations dedicated for LEO satellites. The complexity and size of these ground stations depends on applications. The communication between the satellite and the ground station for scientific missions is usually established on S-band. The main characteristic of this type of stations, is that LEO stations employ tracking antennas to utilize the full capacity, since the satellite flies too fast over the ground station and having too short communication (usually less than 15 minutes) with the appropriate ground station, so the ground station antenna must follow the satellite with the high accuracy.

Thus, even though the communication or the goal of the mission can be accomplished with only one ground satellite station, because of the redundancy and to increase the download data capability, usually more ground stations are used for a single scientific satellite's mission. It is better if there is no overlap between two or more ground stations. The none-overlapping case is presented in Figure 1.8, related to the MOST (Microvariability and Oscillation of Stars) satellite which had ground stations in Canada (Toronto and Vancouver) and in Austria (Vienna) (Northern Lights Software Associates 2003; Keim et al. 2004).

Figure 1.8 shows that the stations in Vancouver and Toronto are overlapped but not with the Vienna ground station; therefore, the stations in Canada and the station in Vienna do not communicate with the LEO satellite at the same time. This increases the download data capability.

The quality of communication depends on the performance of the satellite ground station. Before implementing a satellite ground station, the analysis related to environmental factors must be considered, especially in urban areas. Rain effects, uplink and downlink antenna isolation, intermodulation interference, desensiblization, and analysis related to the contact time duration under

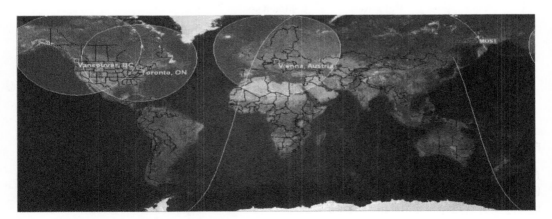

Figure 1.8 Visibility of the MOST satellite.

the low elevation angles are a few aspects treated within this work related to the final decision on design and implementation of the ground station. Most of the satellite services (among them scientific satellites) use frequency bands that are shared with terrestrial services. For systems to coexist, the ITU have specified certain constraints in the transmitted EIRP from the satellite. Such constraints have impact on the design of a ground station. Several trade-offs are necessary in the optimization process up to final design concept of the ground station.

A fundamental parameter in describing a ground station performance is *Figure of Merit* as a ratio of receiving antenna gain to system noise temperature (G/T_S). This Figure of Merit represents the sensitivity of the ground station. A higher value implies a more sensitive station.

The ground station can be considered as two subsystems, the transmit and receive subsystem. From this view, the ground station can be categorized as single antenna or double antenna system. For a single antenna configuration, the antenna is the common element for both subsystems. The transmit subsystem consists of several major components: baseband equipment, modulator, frequency upconverter, high-power amplifier (HPA), and antenna feed system. The receive subsystem behind the antenna feed system uses the low noise amplifier (LNA), frequency downconverter, demodulator, and baseband equipment. A general configuration of a single antenna ground station is shown in Figure 1.9 (Richharia 1999). Signals from the terrestrial network or directly from the user in some applications are fed to a ground station via a suitable interface. The baseband signals are then processed, modulated, and up-converted to the desired satellite transmit frequency. After up-conversion, the signals are amplified by HPA to the required level and transmitted via the antenna.

Signals received through antenna are amplified by LNA, then down-converted to an intermediate frequency (IF), demodulated, and transferred to the terrestrial network via an interface (or directly to the user in some applications). The feed system provides the necessary aperture illumination, introduces the required polarization, and provides the isolation between the transmitted and received signals. Other subsystems such as tracking, control, monitoring, and power supply provide the necessary support. Drive motors enable the ground station's antenna movement to follow the satellite above the ground station. The exact configuration of a ground station depends on applications. To illustrate this, in Figure 1.10 the configuration of the ground station implemented in

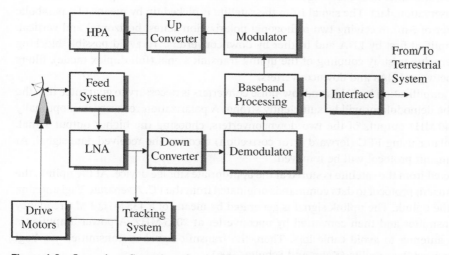

Figure 1.9 General configuration of a single antenna ground station.

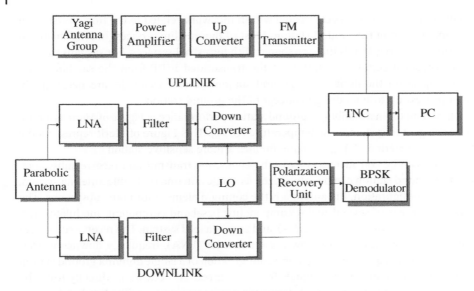

Figure 1.10 Block diagram of the Vienna LEO satellite ground station.

Vienna for communication with MOST satellite is presented, as a double antenna system configuration (Keim and Scholtz 2006). The Vienna ground station system was set up at the Institute for Astronomy of the University of Vienna in cooperation with the Institute of Communications and Radio-Frequency Engineering of the Vienna University of Technology. The Vienna LEO ground station achieved successful two-way communication with the MOST microsatellite on September 30, 2003, until March 2019, when the MOST satellite was deactivated (MOST-A tiny satellite probes the mysteries of universe, 2019). The acronyms in the Figure 1.10 are: BPSK (binary phase shift keying), FM (frequency modulation), LNA, LO (local oscillator), TNC (terminal node controller), and PC (personal computer).

The uplink block is planned to issue commands for: operating the satellite's payload, altitude control subsystem, and other housekeeping functions. The downlink block is responsible for receiving download observation data. The signal from the satellite is picked up by means of a parabolic antenna (diameter of 3 m), receiving two orthogonal polarization states, horizontal and vertical. The signal is amplified first by LNA and further by downconverters. To avoid possible blocking of downconverters due to stray coupling of the uplink transmit signal (full-duplex mode), filters are introduced between LNAs and downconverters.

The two-stage amplification by LNAs and low-noise converters is necessary to guarantee that the signal input to the demodulator will be sufficiently strong. A polarization recovery unit optimally combines the 140 MHz outputs of the two downconverters, choosing the higher output signal. A BPSK demodulator using FEC (forward error correction) recovers the received data signal. At the TNC, the transmit protocol will be removed.

The data collected from the satellite is stored at the appropriate storage device. At the uplink, the TNC adds the transmit protocol to data commands originated from the PC. A separate Yagi antenna group supports the uplink. The uplink signal is generated by means of a 435 MHz FM (frequency modulation) transmitter and then converted by upconverter at 2055 MHz. A power amplifier is placed near the antenna to avoid cable loss. Then, the transmit signal is transmitted via Yagi antenna group toward the satellite (Keim and Scholtz 2006).

1.4 Ground Station Subsystems

The most common ground station subsystems are:

- At the downlink: antenna, low noise amplifier, down converter, and demodulator
- At the uplink: the modulator, upconverter, high-power amplifier and antenna (Figure 1.7)
- Support safety system

These are described next.

1.4.1 Antennas

Because of the large signal attenuation at RF frequencies, the ground station antenna must have high signal gain and must be highly directional to focus power to and from the satellite. Most ground stations use parabolic reflector antennas since such antennas can readily provide high gain and the desirable side lobe characteristics. If the same feed is used for both polarizations, the feeder should separate and combine polarizations in a dual-polarized system. Table 1.1 provides a few typical parameters that should be considered for link budget calculations.

Generally, to avoid combiner (duplexer) loss, at the front-end separate antennas should be used for uplink and downlink, as a double antenna system configuration. The decision on the final antenna system design should be based on required downlink margin. The uplink and downlink antennas must be isolated from each other (Cakaj and Malaric 2007b).

LEO ground stations use tracking antennas, so an antenna mount is also required. The most used mount is *azimuth-elevation mount*, which provides azimuth and elevation angles control. The power control electronics provides the drive signals for the antenna tracking motors. The antenna pointing angle coordinates are precomputed for a satellite pass in the control computer based on the satellite's space orbital parameters. These coordinates are uploaded to the antenna control processor prior to a satellite pass. During a satellite pass, the antenna control processor commands the power electronics module, which positions the antenna in angular alignment with a satellite. The antenna position is updated frequently, depending on the mission's required accuracy (Reisenfeld et al. 2007).

1.4.2 Low Noise Amplifier

The weak signals at the downlink from the satellite are received by the parabolic antenna and then amplified by LNA. Table 1.2 provides some typical technical parameters for link budget calculations and implementation.

Table 1.1 Some of technical antenna's parameters.

Operating frequency	1.0–12	GHz
Gain at operating frequency	35	dBi
Diameter	3	m
Side lobes	−20	dB
Front/back ratio	−25	dB

Table 1.2 Some of technical LNA's parameters.

Operating frequency	2–2.5	GHz
Gain	30	dB
Noise figure	0.6	dB
Operating voltage	12	V

In case of double antenna system, due to coupling between transmitting and receiving antenna, the downlink antenna will also receive the transmitted signal (Cakaj and Malaric 2007b). In this case, filters should be used that efficiently suppress the transmit signal but do not introduce significant loss at the receiving frequency. The isolation between downlink and uplink for the double antenna system should be measured to be kept under allowable limits.

1.4.3 Converters

Up and down converters provide the translation between intermediate frequency (IF), which is typically 70 MHz or 140 MHz, and the actual uplink and downlink frequencies, respectively. The output power level from the downconverter should be sufficient to stimulate a demodulator, and the output power level from the upconverter should be sufficient to drive the power amplifier. In large or professional ground stations, the converters are separate units designed for flexibility, easy for maintenance, and stable operation. Table 1.3 provides typical technical parameters that should be considered for link budget calculations and implementation for converters. The typical downconverter and upconverter block diagrams used in satellite ground stations are presented in Figure 1.11 under a, b respectively (Elbert 1999).

A typical upconverter amplifies the signal to provide adequate gain for the operation of the station equipment. The actual frequency conversion is accomplished in a mixer and LO (local oscillator) combination such as shown in Figure 1.11. A frequency agile upconverter employs a frequency synthesizer to generate the LO so any carrier frequency within the satellite uplink band

Table 1.3 Some of technical converters' parameters.

Input frequency	2232	MHz
Local oscillator frequency	2372	MHz
Output frequency	140	MHz
Gain	32	dB
Noise figure	0.8	dB
Maximum input power	17	dBm
Output power	30	dBm
Spurious signal attenuation	40	dBc
Operating voltage	12	V

(a)

Down converter block diagram

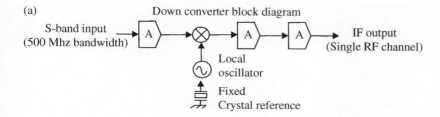

(b)

Up converter block diagram

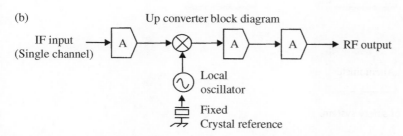

Figure 1.11 Upconverter and downconverter diagrams for satellite ground station.

can be used. Proper filtering is needed to prevent the LO and its harmonics from reaching the uplink path (Elbert 1999).

The downconverter is positioned behind the LNA. The first amplifier stage provides the needed overall gain and reduces the noise contribution of mixer and IF equipment. A synthesizer also must be used to provide agility in the receiving frequency operation (Elbert 1999).

1.4.4   Safety System

LEO satellites establish the lock with the ground station 6–8 times per day for 5–15 minutes at a time. These contacts are established both day and night. Thus, data should be automatically downloaded at the ground station because there will likely be times the station is unattended. The concept of an unattended ground station for LEO missions has been validated by the successful demonstration of a telemetry received terminal in the 2210–2295 MHz band (S-band) that tracked two NASA spacecrafts in LEO (Golshan et al. 1996). Successful demonstrations of the automated, unattended operations of the terminal were conducted with the Solar, Anomalous and Magnetospheric Particle Explorer (SAMPEX) in July and with the Extreme Ultraviolet Explorer (EUVE) in December 1994. Validation of demonstration was accomplished in December 1995 (Losik 1995). A safety/security system must be in place when using an automatic working mode. In principle, this system consists of two elements: storm safety system and visual monitoring system.

Storm safety system is designed to protect the antenna (dish), and the hardware structure from damages due to strong wind (see Figure 1.12). If the wind speed as measured with an anemometer exceeds certain a wind limit (for example of 70 km/h), then the rotator controller will react and the antenna will be brought to an elevation angle of 90° (zenith). In this position, it is protected, since the antenna has the smallest target area for the wind regardless of wind direction (Keim and Scholtz 2006). This is known as *antenna parking position*.

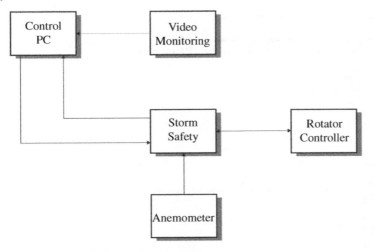

Figure 1.12 Block scheme of safety system.

1.5 Downlink Budget

A communication satellite system enables the communication between a satellite and one or more ground stations. The links used for interconnections should be designed to deliver the information at the destination with acceptable desired signal power level, according to the customer's requirements. A compromise is exercised between the quality of delivered information and practical constrains related to the propagation and the cost/quality of the equipment (Sklar 2005).

The performance of a communication satellite system is mainly expressed in a link budget. Factors that need consideration in a link design are operational frequency, propagation effects, terminal complexity, noise effects, and regulatory requirements. Usually, the costumers of the satellite communication systems in advance define their requirements related to the quality level that the system must fulfill. To satisfy and complete these costumers' requirements, a link power budget should be analyzed and completed. The link budget is a balance sheet of gains and losses; it outlines the detailed participation of transmission and reception resources, noise sources, and all effects throughout the link. The link budget includes both uplink and downlink.

Together with other modeling techniques, the link budget can help to predict equipment, technical risk, performance, and the cost. It is a tool for adjusting the ground station and satellite parameters to satisfy the requirements on the optimal way. An accurate link budget includes many parameters, such as the following (Gordon and Morgan 1993):

- Antenna gain (G)
- Equivalent Isotropic Radiated Power ($EIRP$)
- Free space loss (L_S)
- System noise temperature (T_S)
- Figure of Merit for receiving system (G/T_S)
- Link margin (LM)

1.5.1 Error-Performance

In general, for digital communications, *the error performance* is expressed through E_b/N_0 where E_b is bit energy and N_0 is noise power spectral density (Sklar 2005). Bit energy E_b can be written as:

$$E_b = ST_b \tag{1.1}$$

where S is the signal power and T_b determines the time occupied by a single bit. Also, $N_0 = N/B$ where N is noise power and B is a bandwidth. The bit time T_b is reciprocal with bit rate R_b, and then the error performance can be displayed as:

$$\frac{E_b}{N_0} = \frac{S}{N} \frac{B}{R_b} \tag{1.2}$$

The Eq. (1.2) tells us, that the analysis of error performance of digital communication system can be done through analysis of S/N (signal-to-noise ratio), which is very common from analogue systems. This ratio refers to average signal power and average noise power. The higher the signal-to-noise ratio, under the fixed bandwidth and the fixed bit rate, the better is the energy bit over noise density ratio.

The ratio S/N can degrade by two reasons:

- Through the decrease (loss) of the desired signal power (S)
- Through the increase of the noise power (N)

Furthermore, we will call these degradations, respectively, *loss* and *noise*. Losses are:

- *Free space loss* is a decrease of the wave's power simply as a function of distance. For a satellite communication link, the free space loss is the largest loss because of the long distance between transmitting and receiving terminals.
- *Atmospheric loss* includes all effects of atmosphere with the influence on decrease of the signal power.
- *Polarization loss* is a loss of signal due to any polarization mismatch between the transmitting and receiving antennas.
- *Pointing loss* is a loss of signal when either the transmitting antenna or receiving antenna is imperfectly pointed.
- *Noise* has several sources along signal's route such are: thermal, atmospheric, galaxy, and interference from other sources.

1.5.2 Received Signal Power

The main purpose of the link budget is to verify that the communication system will operate according to the predicted-designed plan. This means that the error performance will meet the specifications. In radio communication systems, the carrier power is propagated from the transmitter using transmitting antenna, which will then be received through receiving antenna. The development of the fundamental relationship between transmitted and received power usually begins with the assumption of an omnidirectional RF (radio frequency) source, transmitting uniformly over 4π steradians. Such an ideal source, called an isotropic radiator, is illustrated in Figure 1.13 (Sklar 2005).

The power density $p(d)$ on a hypothetical sphere at the distance d from the source is related to the transmitted power P_t as:

$$p(d) = \frac{P_t}{4\pi d^2} \tag{1.3}$$

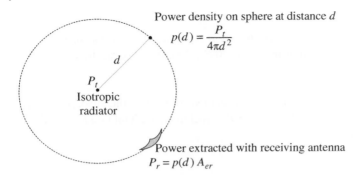

Figure 1.13 Received power from an isotropic antenna.

where $4\pi d^2$ is the sphere's area. The extracted power with the receiving antenna will be:

$$P_r = p(d)A_{er} \tag{1.4}$$

$$P_r = \frac{P_t A_{er}}{4\pi d^2} \tag{1.5}$$

where P_r is received power and A_{er} is the absorption cross section (effective area) of the receiving antenna. The parameters without index are general for both. An antenna's effective area A_{er} and physical area A_p are related by an efficiency parameter as:

$$A_{er} = \eta A_p \tag{1.6}$$

which clarifies that total incident power is not extracted. Values of η are in range from 0.4 to 0.8 but the most common value is $\eta = 0.55$ (Sklar 2005). Further, all parameters related to transmission terminal will be designated by index t and to receiving terminal by index r.

The effective radiated power or *equivalent isotropic radiated power* (EIRP) with respect to an isotropic source is defined as a product of transmitted power P_t and the gain of the transmitting antenna G_t, as follows:

$$EIRP = P_t G_t \tag{1.7}$$

However, for the more general case in which the transmitter has an antenna gain relative to an isotropic antenna, the P_t in Eq. (1.5) can be replaced with $P_t G_t$ or with EIRP. Based on Eq. (1.7), then, Eq. (1.5) becomes

$$P_r = EIRP \frac{A_{er}}{4\pi d^2} \tag{1.8}$$

The relationship between antenna gain G and antenna effective area A_{er} is:

$$G = \frac{4\pi A_{er}}{\lambda^2} \tag{1.9}$$

where λ the wavelength of the carrier is. Wavelength λ and frequency f are related by Eq. (1.10),

$$\lambda = \frac{c}{f} \tag{1.10}$$

where c is the light's velocity (considered as $c = 3 \cdot 10^8 \text{m/s}$).

The expression in Eq. (1.9) is similar for both transmitting and receiving antennas. The *reciprocity theorem* states that for a given antenna and carrier wavelength, the transmitting and receiving gains are identical. All discussion is done under assumption of an isotropic source, and then the antenna gain is $G = 1$. Putting this at Eq. (1.9) yields

$$A_{er} = \frac{\lambda^2}{4\pi} \tag{1.11}$$

To find the received power P_r when the receiving antenna is isotropic, we substitute Eq. (1.11) at Eq. (1.8) to get

$$P_r = \frac{EIRP}{(4\pi d/\lambda)^2} = \frac{EIRP}{L_S} \tag{1.12}$$

where collection of terms $(4\pi d/\lambda)^2$ is known as *free space loss* designated by L_S.

$$L_S = \left(\frac{4\pi d}{\lambda}\right)^2 = \left(\frac{4\pi d f}{c}\right)^2 \tag{1.13}$$

For the more general case, when the receiving antenna is not isotropic, this means it has an antenna gain G_r. The Eq. (1.12) becomes more general, as:

$$P_r = \frac{(EIRP)G_r\lambda^2}{(4\pi d)^2} = \frac{(EIRP)G_r}{L_S} \tag{1.14}$$

The Eq. (1.14) shows that the received power depends on *effective radiated power, free space loss,* and *receiving antenna gain.* Usually, all above-mentioned parameters are expressed in (dB); however, the *equation* expressed in (dB) is:

$$P_r = EIRP - L_S + G_r \tag{1.15}$$

Free space loss is the greatest loss in transmitted power due to the long distance between the satellite and ground station. The free space loss L_S strongly depends on distance d and frequency f, as presented in in Figure 1.14.

The diagram is presented for frequencies of 2 GHz, 8 GHz, and 12 GHz and for distances from 1,000 km up to 36,000 km, including the highs related to LEO, MEO, and GEO orbits. Figure 1.14 shows that free space loss increases by both frequency and the distance. This has a large impact on transmitted signal power.

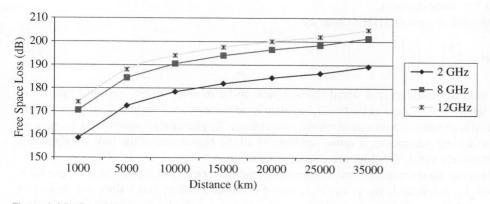

Figure 1.14 Free space loss.

1.5.3 Link Budget Analyses

Evaluating system performance, the previous analysis has shown that the quantity of the highest interest is $S/N(SNR)$, or signal-to-noise ratio. In satellite communication systems, mostly the carrier is frequency or phase modulated (frequency shift keying (FSK) or phase shift keying (PSK)), having a constant envelope. This means that the signal power and carrier power are the same; thus, the signal power will keep notation S. Then the ratio S/N can be obtained from Eq. (1.12), (1.14), divided by N as:

$$\frac{S}{N} = \frac{EIRP(G_r/N)}{L_S} \tag{1.16}$$

For digital communications systems, Eq. (1.16) is used to express this ratio by noise power spectral density. Thus, applying $N_0 = kT_s$ yields:

$$\frac{S}{N_0} = \frac{EIRP(G_r/T_S)}{kL_S} \tag{1.17}$$

where T_S is system noise temperature, representing the noise radiated into antenna and thermal noise generated by the receiving system. L_S is free space loss. The system effective noise temperature T_S is a parameter that models the effects of various noise sources. The ratio G_r/T_S is known as a Figure of Merit. The Eq. (1.17) expressed in (dB) follows:

$$\frac{S}{N_0} = EIRP - L_S + G_r/T_S + 228.6 \tag{1.18}$$

In this equation the value of $228.6 dBW/HzK$ yields from Boltzmann's constant. The signal to noise power expressed as logarithmic equation is:

$$\frac{S}{N}(dB) = \frac{S}{N_0}(dB) - B(dB) \tag{1.19}$$

Within link budget calculations, the atmospheric losses and other degradation factors that affect the received power must be considered. Then, if we introduce the term L_0, which represents all other loss factors, and apply it at Eq. (1.18), expressed in dB, we'll get:

$$\frac{S}{N_0} = EIRP - L_S - L_0 + G_r/T_S + 228.6 \tag{1.20}$$

known as the *range equation*.

The downlink margin (DM) is defined as:

$$DM = \left[\frac{S}{N}\right]_r - \left[\frac{S}{N}\right]_{rqd} \tag{1.21}$$

where r indicates the expected signal-to-noise ratio to be received at receiver, and rqd means required signal to noise ratio by customer, based on in advance defined performance. So, a positive value of DM is an indication of a good system performance. To guarantee a positive link margin, we must trade among parameters of range equation. If all the parameters of the link are rigorously treated (the worst case), high link margin is not mandatory.

Thus, in principle the needed items to calculate signal power over noise power density are $EIRP$, G_r/T_S, and (L_S, L_0). $EIRP$ is the power transmitted from the satellite, and it does not depend on environmental factors. G_r is receiving antenna gain. Other loss includes atmospheric, polarization, and pointing loss. The atmospheric impairments are analyzed latter on, and the pointing loss can be avoided by accurate pointing equipment. Of further interest remain free space loss and system temperature.

Since LEO satellites move too fast over the Earth, the satellite's slant path range (distance in between the ground station and the satellite) varies over the time, on dependence of the elevation angle, so, the signal toward the ground station is faced with different distances from the satellite to the ground station, and consequently different free space loss under different elevation. The variation on free space loss under different elevation impacts signal-to-noise spectral density ratio, and consequently the receiving ground station performance. The following equation expresses the dependence of free space loss on the elevation angle under which one the ground station sees the satellite:

$$L_s(\varepsilon_0) = \left(\frac{4\pi f}{c}\right)^2 d^2(\varepsilon_0) \tag{1.22}$$

The next sections of this chapter will discuss the system temperature, followed by satellite ground station geometry.

1.6 Figure of Merit and System Noise Temperature

For the satellite communication system, the performance of the receiving system, or known as the downlink performance, what is the subject of this book, is commonly defined through a receiving system Figure of Merit as G/T_S, where:

$$T_S = T_A + T_{comp} \tag{1.23}$$

Here, G is receiving antenna gain, T_S is receiving system noise temperature, T_A is antenna noise temperature and T_{comp} is composite noise temperature of the receiving system, including lines and equipment. The composite temperature depends exclusively on parameters of technical equipment and of interconnection lines characteristics. Otherwise, the antenna temperature T_A depends on external environment factors also including the sky background represented by its sky noise temperature denoted as T_C.

Schematically, the satellite ground station receiving system and the environment concept is presented in Figure 1.15.

Unwanted noise power is, in part, injected via antenna ($kT_A B$) and part is generated internally ($kT_{comp}B$) by line loss and equipment. k is Boltzmann's constant and B is system bandwidth. T_c represents the sky noise temperature, T_m is medium temperature, and A is medium attenuation (Saunders 1993). Further will be discussed antenna noise temperature and composite noise temperature, both components of system temperature expressed by Eq. (1.23).

Different noise sources (natural, man-made, or interferences) at surrounding environment are present in the front of the receiving antenna of the satellite ground station system. The antenna will pick up part of this noise. The picked-up noise power from these external sources is given as $kT_A B$, where T_A is *antenna noise temperature*. The picked-up noise power depends on where the antenna is looking at. The antenna noise temperature is a measure of the effective temperature integrated over the entire antenna pattern.

In general, the total antenna noise power will be made up by the various sources whose temperature will vary with the space angle of observation (θ, ϕ). This power will be picked up by the

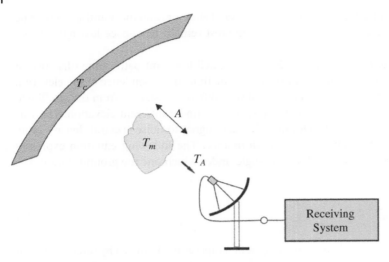

Figure 1.15 Satellite ground station and environment concept.

antenna, which power pattern is $F(\theta, \phi)$ (Saunders 1993; Nikolova 2002). Generally, antenna temperature can be expressed by:

$$T_A = \frac{1}{\Omega_A} \oint_{4\pi} \oint_{4\pi} F(\theta, \phi) \cdot T_C(\theta, \phi) d\Omega \qquad (1.24)$$

where $T_C(\theta, \phi)$ is sky noise source temperature and Ω_A is antenna beam angle under which antenna sees this source. Sky noise temperature is generated from different sky sources (cosmic radiation, Sun, Moon, stars etc.), and the antenna adds internal noise to the receiving system; both degrade the downlink's performance.

The cosmic background radiation (D), presented in Figure 1.16, is independent of frequency and appears everywhere in the sky at the temperature of around 3–10 K. The galactic noise temperature by stars is also presented. The range of these variations is indicated as region (C). This noise decreases rapidly with frequency (Saunders 1993).

Considering that the entire antenna pattern (beam) sees a sky noise source under the same temperature conditions, then $T_C(\theta, \phi) = const = T_C$. This assumption is valid when the solid angle Ω_C

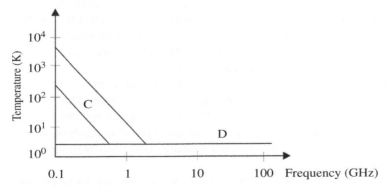

Figure 1.16 Sky noise temperature sources.

subtended by the noise source is much larger than the antenna solid angle Ω_A, which is the case for the receiving antennas at LEO satellite ground stations. Next, the antenna itself is considered lossless; it does not generate noise itself. Then, Eq. (1.24) becomes:

$$T_A = \frac{T_C}{\Omega_A} \oint_{4\pi} F(\theta, \phi) d\Omega \tag{1.25}$$

Since:

$$T_C(\theta, \phi) = const = T_C \tag{1.26}$$

and

$$\oint_{4\pi} F(\theta, \phi) d\Omega = \Omega_A \tag{1.27}$$

In case of $\Omega_A \ll \Omega_C$, yields out:

$$T_A = T_C \tag{1.28}$$

This means, that the antenna temperature T_A is equal to the sky source temperature T_C. From Figure 1.16 we see that for LEO satellites operating at 2 GHz band, sky noise temperature does not depend on galactic noise (region C), so it remains as constant within LEO satellites operating frequency range.

When an atmospheric absorptive process takes place (see Figure 1.15), the absorption increases the antenna noise temperature. If we consider the total cosmic temperature as T_C, the absorptive process (rain) medium temperature as T_m and the attenuation due the absorptive process as A, then the total antenna noise temperature T_A of the receiving satellite ground station is (Saunders 1993):

$$T_A = T_m \left(1 - 10^{-A/10}\right) + T_C 10^{-A/10} \tag{1.29}$$

where, typically T_m is 275 K to 290 K for rain. So, finally three components, two temperatures T_m, T_C and medium attenuation A will determine antenna noise temperature T_A.

Let us move to the second sum component, respectively, to the composite noise temperature T_{comp} analysis. Figure 1.17 presents the first stage of a satellite receiving system, including the elements where the loss and noise will have the primary role on S/N degradation. These elements are the antenna, line, and equipment (as the front-end device is the LNA). For calculations of the composite noise temperature equipment and line will be treated.

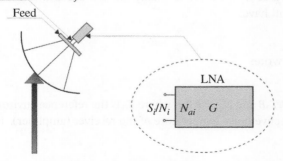

Feed connection cable (L_f)

Feed

LNA

S_i/N_i N_{ai} G

Figure 1.17 The first stage of a satellite receiving system.

Let us generally touch the *thermal noise*, where one degrades the signal-to-noise power ratio at the receiver. *Thermal noise* is caused because of thermal motion of electrons at physical temperature T. These motions of electrons generate the electromagnetic radiation. Part of this radiation in microwave frequencies will be present at the receiving system. The noise power spectral density N_0 is constant at all frequencies, known as *white noise*.

The noise power N within a bandwidth B is:

$$N = kTB \tag{1.30}$$

T is temperature in Kelvin (K), $k = 1.38 \cdot 10^{-23}$W/HzK is Boltzmann's constant.

The noise power spectral density N_0 (noise power within a bandwidth $B = 1$Hz) is:

$$N_0 = N/B = kT \tag{1.31}$$

From Eq. (1.31), the noise power density of thermal noise depends on the ambient temperature of the source. This leads to the useful concept of an *effective noise temperature* for noise sources that are not necessarily thermal in origin (e.g., galactic, atmospheric), which can be introduced into the receiving antenna. The noise power of such sources can be expressed separately through the effective noise temperature of a hypothetical thermal noise source power. The total effect of such noise sources, including external and internal, will be expressed through *system temperature*.

Noise figure is a parameter that expresses the noisiness of two port networks or devices (such as LNA). Noise figure F relates the S/N at the input of a network or device to the S/N at the output of the network or device. Noise figure of the preamplifier (LNA) at the receiving system shown in Figure 1.17 is defined as:

$$F = \frac{(S/N)_{in}}{(S/N)_{out}} = \frac{S_i/N_i}{GS_i/G(N_i + N_{ai})} \tag{1.32}$$

where S_i is signal power at the amplifier input port, N_i is noise power at the amplifier input port, N_{ai} is amplifier's internal noise referred to the input port and G is amplifier's gain. The Eq. (1.32) can be reduced to:

$$F = \frac{N_i + N_{ai}}{N_i} = 1 + \frac{N_{ai}}{N_i} \tag{1.33}$$

An ideal amplifier with no internal noise ($N_{ai} = 0$) has noise figure $F = 1$ or $F\,(dB) = 0\,dB$. For the concept of the noise figure to have utility a value of N_i must be defined as a reference. The noise figure of any device then represents the measure compared with the reference value. In 1944, Fries suggested the noise figure should be defined for a noise source at a reference temperature of $T_0 = 290$K (Saunders 1993). From Eq. (1.31), it can be seen that the noise power spectral density from any source is characterized by appropriate noise temperature. The value of 290K is chosen as a reference because it is reasonable source temperature for many links. If substitutes $T = 290$K at Eq. (1.31) and expresses it in (dB) will have:

$$N_0 = -204\ (dBW/Hz) \tag{1.34}$$

By rearranging Eq. (1.33), we can write:

$$N_{ai} = (F-1)N_i \tag{1.35}$$

At Eq. (1.35) can be replaced $N_i = kT_0B$ and $N_{ai} = kT_RB$ where T_0 is the reference environmental temperature and T_R is called the effective noise temperature of the receiver (amplifier). Then the Eq. (1.35) becomes:

$$kT_RB = (F-1)kT_0B \tag{1.36}$$

$$T_R = (F-1)T_0 \tag{1.37}$$

Finally, for $T_0 = 290K$ we will get:

$$T_R = (F - 1) \cdot 290K \tag{1.38}$$

or, typically for the LNA in Figure 1.17, is:

$$T_{LNA} = (F_{LNA} - 1) \cdot 290K \tag{1.39}$$

Finally, the LNA noisiness expressed through noise figure F_{LNA}, is represented through LNA effective noise temperature. Thus, the noisiness of an amplifier is manifested through the noise temperature, what will further be seen how this approach mathematically simplifies the Figure of Merit calculations for LEO satellite ground station. The last equation tells us that the noisiness of an amplifier can be modeled as it was caused by a noise source, operating at some effective temperature $T_R(T_{LNA})$.

The next element in the Figure 1.17 is the line that connects the feed and LNA. Analyzing LNA, it was seen that S/N degradation resulted from injecting additional (amplifier's) noise into the link. However, in the case of line loss, the S/N degradation results from the signal being attenuated by the transmission line. Considering the line as a network where the line is matched with characteristic impedance at the source and at the load, we can define the power loss as:

$$L = \frac{P_{in}}{P_{out}} \tag{1.40}$$

Considering the network gain as $G = 1/L$, where L is the line loss, and applying the same methodology as for amplifier, we can find out that the *effective noise temperature for line loss* is:

$$T_L = (L - 1) \cdot 290K \tag{1.41}$$

or, typically for the case in Figure 1.17, related to the interconnection line of the antenna feeder with LNA it is:

$$T_{Lf} = (L_f - 1) \cdot 290K \tag{1.42}$$

We have analyzed the noise temperature for a single device and for a single interconnection line, but in the real-world systems, there are more components interlinked to each other through more lines, linked as a chain (known as series or cascade interconnection). Each of them affects the system noise. The whole impact of devices and links within a system is defined as a *composite noise temperature*, including the effect of all equipment and lines. To analyze composite effect of all system components, we will first consider two networks interconnected in series with noise figures respectively F_1 and F_2 presented in Figure 1.18a (Sklar 2005).

Based on Eq. (1.32) and Eq. (1.33) as a definition for noise figure and simply mathematical operations, we can find out that the composite noise figure for these two networks is:

$$F_{comp} = F_1 + \frac{F_2 - 1}{G_1} \tag{1.43}$$

(a)

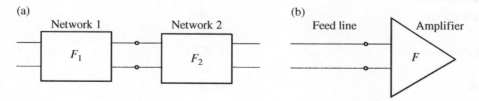

(b)

Figure 1.18 Networks connected in series.

where the G_1 is the gain of the Network 1. For the typical case, from the Figure 1.18b, are $F_1 = L_f$, $G_1 = G_f = 1/L_f$ and $F_2 = F_{LNA}$. Substituting these at Eq. (1.43) will get F_{comp} as:

$$F_{comp} = L_f + L_f(F_{LNA} - 1) = L_f F_{LNA} \tag{1.44}$$

Applying relation between F_{comp} and T_{comp} as:

$$T_{comp} = (F_{comp} - 1)290K \tag{1.45}$$

yields out:

$$T_{comp} = (L_f F_{LNA} - 1) \cdot 290K \tag{1.46}$$

The composite temperature can be displayed also as a function of effective noise temperature of the preamplifier (T_{LNA}) and of effective noise temperature of line loss (T_{Lf}) as follows:

$$T_{comp} = [(L_f - 1) + L_f(F_{LNA} - 1)] \cdot 290K = T_{Lf} + \frac{1}{G_f} T_{LNA} \tag{1.47}$$

To find out the composite noise figure F_{comp} and composite noise temperature T_{comp} of n networks, which are connected in cascades and characterized with F_n and G_n, the following equations apply:

$$F_{comp} = F_1 + \frac{F_2 - 1}{G_1} + \frac{F_3 - 1}{G_1 G_2} + ... + \frac{F_n - 1}{G_1 G_2 ... G_{n-1}} \tag{1.48}$$

$$T_{comp} = T_1 + \frac{T_2}{G_1} + \frac{T_3}{G_1 G_2} + ... + \frac{T_n}{G_1 G_2 ... G_{n-1}} \tag{1.49}$$

Finally, for the general case, to determine the Figure of Merit (G/T_S) for the satellite ground station, apply Eq. (1.23), (1.29), and Eq. (1.49), with respective parameters at each equation. Usually, the link budget is presented in the tabulated format, for both uplink and downlink with all appropriate parameters included the order of steps of their calculation given under Tables 1.4 and 1.5. In the following chapters practical calculations are provided.

Table 1.4 Downlink budget.

Transmit power	**dBW**
Loss	dB
Antenna gain	dBi
EIRP	dBW
Total propagation loss	dB
Received isotropic power	dBW
Antenna gain	dBi
System noise temperature	dBK
Figure of Merit (G/T_S)	dB/K
S/N_0	dB
Receiver bandwidth	dBHz
S/N	dB
Required S/N	dB
Downlink margin	dB

Table 1.5 Uplink budget.

Transmit power	dBW
Loss	dB
Antenna gain	dBi
EIRP	dBW
Total propagation loss	dB
Received isotropic power	dBW
Antenna gain	dBi
System noise temperature	dBK
Figure of Merit (G/T_S)	dB/K
S/N_0	dB
Receiver bandwidth	dBHz
S/N	dB
Required S/N	dB
Uplink margin	dB

1.7 Satellite and Ground Station Geometry

Theoretically, the position of the orbit is fixed in space, as for further analysis it is considered. The position of the satellite in space is determined by space orbital elements, in this order, the position of the orbital plane in space (orbit lays on orbital plane) is defined in reference to the Vernal equinox and equatorial plane, further the position of the orbit within its orbital plane is determined by argument of perigee (referred to line of nodes), and further the position of the satellite within its orbit is determined with its true anomaly angle (referred to the perigee) (Cakaj et al. 2007c).

As the satellite orbits, the Earth where the ground station sits rotates as well. Because of this, the distance between the ground station and the satellite changes over time, for LEOs and MEOs. Since LEOs move faster, these variations in distance between the satellite and ground station happen faster than under the case with MEOs. This is not the case with GEOs, since the orbit is synchronized with the Earth's rotations and thus the distance between the ground station and the satellite will not change under normal conditions.

The main goal of the satellite systems is establishing the communication between the satellite and the ground station. The location of the ground station is usually given in terms of geographical coordinates defined as *latitude* and *longitude*. Thus, for link budget calculation (which enables the communication by providing sufficient signal to noise level), correlation between the satellite position and the ground station location should be established, and mathematically expressed. This in fact brings the problem on finding out the slant range in between the ground station and the satellite, for the look angles under which the satellite is seen from the ground station. But how is the satellite seen from the ground station?

The position of the satellite within its orbit considered from the ground station point of view can be defined by *azimuth* and *elevation* angles. The horizon plane for a given ground station is depicted in Figure 1.19, to define the concepts of *azimuth* and *elevation*.

The *azimuth* (A_z) is the angle of the direction of the satellite, measured in the horizon plane from geographical north in clockwise direction. The *elevation* (ε_0) is the angle between a satellite and the

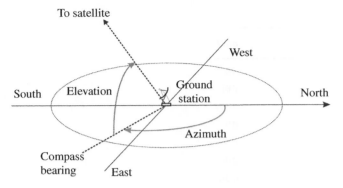

Figure 1.19 Azimuth and elevation.

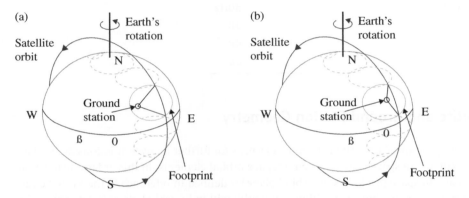

Figure 1.20 Satellite passes for an Earth rotation angle of β per orbit.

observer's (ground station) horizon plane. These two quantities are of primary interest for pointing a tracking antenna at LEO ground station to the satellite.

LEO satellites, being closer to the Earth, orbit several times daily around the Earth. Because of the Earth's motion around its north–south axis the satellite passes related to the determined ground station change from pass to pass. The orbital plane is in principle fixed and defined by orbital parameters, so the orbit keeps its position unchanged, but because of Earth's rotation around its N-S axis for angle β the ground station changes the position relatively to orbital plane, so the pointing (look angles) from the ground station to the satellite are not identical at both passes, for the same satellite's orbital position in space (Figure 1.20a,b) (Roddy 2006). This is illustrated in Figure 1.20.

To further clarify the correlation between the satellite and the ground station location, more exactly to find out the distance between the ground station and the satellite, in Figure 1.21 the position of a satellite within inclined orbital plane (LEO) with respect to the ground station is presented.

In Figure 1.21 both the satellite radius vector $\vec{r}$ and the ground station radius vector $\overrightarrow{R_E}$ for any position of the satellite and the ground stations are known. From the Figure 1.21 yields:

$$\vec{d} = \vec{r} - \overrightarrow{R_E} \tag{1.50}$$

$\vec{d}$ is the satellite to ground station range vector.

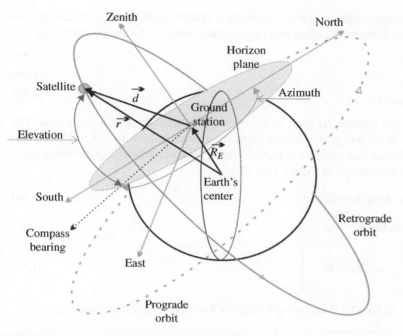

Figure 1.21 Satellite position related to the ground station.

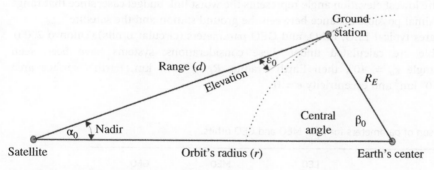

Figure 1.22 Ground station geometry.

Through this approach, the satellite to ground station range vector $\vec{d}$ is transformed to topocentric-horizon system, which enables the look angles and the range from the ground station to the satellite to be calculated.

The triangle from the space view in Figure 1.21 is brought on plane, as a basic geometry between a satellite and ground station and depicted in Figure 1.22 (Gordon and Morgan 1993).

The two points indicate the satellite and ground station, and then the third is the Earth's center. Two sides of this triangle are usually known (the distance from the ground station to the Earth's center, $R_E = 6378 \times 10^3$ m and the distance from the satellite to Earth's center-orbital radius). The angle under which the satellite sees the ground station is called *nadir angle*. There are four variables in this triangle: ε_0 is elevation angle, α_0 is nadir angle, β_0 is central angle, and d is slant range. As soon as two quantities are known, the other two can be found with the following equations:

$$\varepsilon_0 + \alpha_0 + \beta_0 = 90 \tag{1.51}$$

$$d\cos\varepsilon_0 = r\sin\beta_0 \tag{1.52}$$

$$d\sin\alpha_0 = R_E\sin\beta_0 \tag{1.53}$$

Sometimes one of the four parameters must be calculated in terms of any of the other three. The most needed parameter is the range d (distance from the ground station to the satellite). This parameter will be used during link budget calculation, and it is expressed through elevation angle ε_0. Applying cosines law for triangle at Figure 1.22. yields:

$$r^2 = R_E^2 + d^2 - 2R_E d\cos(90 + \varepsilon_0) \tag{1.54}$$

Solving Eq. (1.54) by d, yields:

$$d = R_E\left[\sqrt{\left(\frac{r}{R_E}\right)^2 - \cos^2\varepsilon_0} - \sin\varepsilon_0\right] \tag{1.55}$$

Substituting, $r = H + R_E$ at Eq. (1.55) we will get range as function of elevation angle ε_0 as:

$$d = R_E\left[\sqrt{\left(\frac{H + R_E}{R_E}\right)^2 - \cos^2\varepsilon_0} - \sin\varepsilon_0\right] \tag{1.56}$$

The range under the lowest elevation angle represents the worst link budget case, since that range represents the maximal possible distance between the ground station and the satellite.

Table 1.6 compares typical LEO, MEO, and GEO parameters (circular orbits) (Difonzo 2000). Data on this table are calculated under these considerations: systems have been seen under elevation angle $\varepsilon_0 = 10°$, then Earth's radius $R_E = 6378$ km, Earth's surface area $A_{EARTH} = 511.2 \cdot 10^6$ km^2 and eccentricity $e = 0$.

Table 1.6 Comparison of parameters for LEO, MEO and GEO orbits.

Orbit	LEO	MEO	GEO
System	Iridium	ICO	INTELSAT
Inclination i (°)	86.4	45	0
Altitude H (km)	780	10 400	35 786
Semi major axis a (km)	7159	16 778	42 164
Orbit period (min)	100.5	360.5	1436.1
$(H + R_E)/R_E$	1.12	2.63	6.61
Earth central angle β_0 (°)	18.6	58.0	71.4
Nadir angle α_0 (°)	61.3	22	8.6
Slant range d (km)	2325	14 450	40 586
One-way time delay (ms)	2.6	51.8	139.1
Fraction of covered Earth's area	0.026	0.235	0.34

1.8 LEO MOST Satellite and Ground Stations

In 1997, anticipating new microsatellite altitude control technology, a team of scientists proposed a project to the Canadian Space Agency (CSA) to obtain astronomical photometry of unprecedented precision from a microsatellite. The following year, this project was approved and defined as: Microvariability and Oscillations of Stars (MOST). The MOST astronomy mission under the CSA Small Payloads Programs is the first Canada's space science microsatellite and the first space telescope. The MOST science team used the satellite to conduct long-duration stellar photometry observations in space. These stellar ultra-precise photometry observations are accomplished using a 15 cm aperture optical telescope mounted in a small suitcase sized satellite bus (65 cm × 65 cm × 30 cm; 60 kg) as given in Figure 1.23 (MOST inside, 2019). The high photometric precise optical telescope was developed by the University of British Columbia, and the high-performance altitude control system was provided by Dynacon Enterprises Limited, Canada (Zee and Stibrany 2002). Simplified, the goals of the mission were to analyze the inner structure of stars, set a lower limit to the age of the universe and to search for Exoplanets.

The project MOST consists of a LEO MOST satellite and three ground stations in: Vancouver, Toronto, and Vienna. The satellite link operates on 2GHz band.

The MOST satellite carries instruments designed to observe stars within the satellites CVZ (continuous viewing zone) by measuring tiny light variations undetectable from Earth. MOST is designed to detect variations in the brightness of stars with high precision. This will allow the MOST science team to translate these variations of nearby stars into information about their internal structures and ages, through a technique known as a stellar seismology. In addition, MOST can be used to detect orbiting Exoplanets.[1] Figure 1.24 shows the MOST satellite concept. The MOST satellite is launched in the LEO.

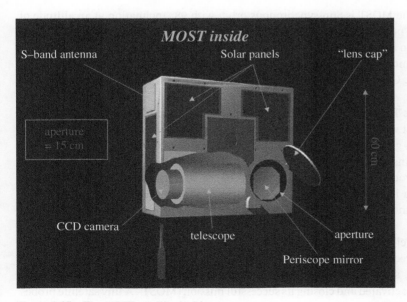

Figure 1.23 The MOST satellite inside.

---

1 An exoplanet is an extrasolar planet that orbits a star other than Earth's Sun.http://http://www.answers.com/topic/extrasolar-planet.

Continuous
viewing zone

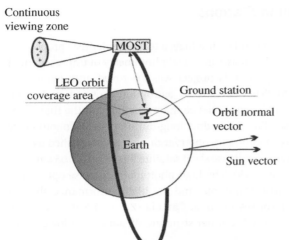

Figure 1.24 MOST satellite concept.

Table 1.7 MOST orbital elements.

Apogee	Perigee	Inclination	Orbital period
839.6 km	825.9 km	98.72 °	6085 s

The baseline orbit of the MOST is Sun-synchronous orbit; with 98° inclination at an altitude of around 820 km. Table 1.7 presents the main MOST orbital elements (Walker et al. 2003). MOST was successfully launched into its specified orbit on June 30, 2003, from the Khrunichev State Research and Production Space Centre in Plesetsk, Russia. MOST was injected into a low orbit with approximate altitude of 820 km. Contact with the satellite was established during it's the first pass over Toronto.

Inclination given in the Table 1.7 makes obvious Sun synchronization. The MOST satellite has a line-of-sight radio contact with each ground station 6–8 times per day when it is above the ground station. A pass over each ground station will last about 5–15 minutes (Carroll et al. 1998).

The project MOST consists of a LEO satellite and three ground stations, one of them in Vienna. The Vienna ground station system was set up at the Institute for Astronomy of the University of Vienna in cooperation with the Institute of Communications and Radio Frequency Engineering of the Vienna University of Technology, given in Figure 1.25 (Cakaj and Malaric 2007a).

In 2008, the MOST Satellite Project Team won the Canadian Aeronautics and Space Institute's Alouette Award, which recognizes outstanding contributions to advancement in Canadian space technology, applications, science, or engineering. The satellite MOST, by March 2019, having 66 487 revolutions was deactivated and closing its successful mission (MOST– A tiny satellite probes the mysteries of universe, 2019).

The author operated with the ground station in Vienna for a period of six months, so a lot of analyses and measurement given in this book stem from this project.

Figure 1.25 Vienna satellite ground station.

References

Botta, A. and Pescape, A. (2013). New generation satellite broadband internet service: should ADSL and 3G worry. TMA, co-located with IEEE INFOCOM 2013, Turin, Italy, April 2013.

Cakaj, S. (2021). The parameters comparison of the "Starlink" LEO satellites constellation for different orbital shells. *Frontiers in Communications and Networks.* 2: 643095. https://doi.org/10.3389/frcmn.2021.643095.

Cakaj, S. and Malaric, K. (2007a). Rigorous analysis on performance of LEO satellite ground station in urban environment. *International Journal of Satellite Communications and Networking* 25 (6): 619–643.

Cakaj, S. and Malaric, K. (2007b). *Isolation Measurement between Uplink and Downlink Antennas at Low Earth Orbiting Satellite Ground Station*, 24–26. Dubrovnik, Croatia: IEEE 19th International Conference on Applied Electromagnetics and Communications – ICECom.

Cakaj, S., Keim, W., and Malaric, K. (2007c). *Communications Duration with Low Earth Orbiting Satellites*, 85–88. Montreal: 4th IASTED International Conference on Antennas, Radar and Wave Propagation, IASTED, ARP.

Cakaj, S., Fischer, M., and Schotlz, L.A. (2009). *Sun Synchronization of Low Earth Orbits (LEO) through Inclination Angle*, 155–161. Austria: Innsbruck: Proceedings of the 28th IASTED International Conference on Modelling, Identification and Control, MIC.

Cakaj, S., Fitzmaurice, M., Reich, J., and Foster, E. (2010a). *Simulation of Local User Terminal Implementation for Low Earth Orbiting (LEO) Search and Rescue Satellites*. The Second International Conference on Advances in Satellite and Space Communications SPACOMM 2010, IARIA, 140–145. Athens, Greece: IEEE.

Cakaj, S., Kamo, B., Lala, A. et al. (2015a). The velocity increment for Hohmann coplanar transfer from different low earth orbits. *Frontiers in Aerospace Engineering* 4 (1): 35–41. https://doi.org/10.12783/fae.2015.0401.04.

Carroll, A.K., Zee, E.Z., and Matthews, J. (1998). The MOST Microsatellite mission: Canada's first space telescope. In: *12th Annual USU/AIAA Conference on Small Satellites*, Logan Utah, 1–19.

Cochetti, R. (2015). Low earth orbit (LEO) mobile satellite communications systems. In: *Mobile Satellite Communications Handbook*, 119–155. Hoboken, NJ: Wiley Telecom.

Constellation One Web, 2022, https://satellitemap.space/?constellation=oneweb.

De Selding, B.P. (2015). *Virgin, Qualcomm Invest in OneWeb Satellite Internet Venture*. Paris: SpaceNews.

Difonzo, F.D. (2000). *Satellite and Aerospace, the Electrical Engineering Handbook, Chapter 74*. Boca Raton: CRC Press LLC.

Elbert, B. (1999). *Introduction to Satellite Communication*. Norwood: Artech House Inc.

Foust, J. (2018) Data from Telesat to announce manufacturing plans for LEO constellation in coming months. SpaceNews. https://spacenews.com/telesat-to-announce-manufacturing-plans-for-leo-constellation-in-coming-months/ (Accessed January 17, 2021).

Garner, P., Cooke, D., and Haslehurst, A. (2009). *Development of a Scalable Payload Downlink Chain for Highly Agile Earth Observation Missions in Low Earth Orbit*, 529–534. Istanbul, Turkey: 4th International Conference Recent Advances in Space Technologies.

Golshan, N., Raferly, W., Ruggier, C., Wilhelm, M., Hagerty, B., Stockett, M., Cuccihissi, J., McWatters, D. (1996). *Low Earth orbiter demonstration terminal*, TDA Progress report 42–125, pp. 1–15, January – March 1996. http://ipnpr.jpl.nasa.gov/progress_report/42-125/125G.pdf (Sep. 2007).

Gordon, D.G. and Morgan, L.W. (1993). *Principles of Communication Satellites*. Sussex: Wiley.

Keim, W. and Scholtz, L.A. (2006). *Performance and Reliability Evaluation of the S-band, at Vienna Satellite Ground Station*, 5. Palma de Mallorca, Spain: Talk, IASTED, International Conference on Communication System and Networks.

Keim, W., Kudielka, V., and Scholtz, L.A. (2004). A scientific satellite ground station for an urban environment. In: *International Conference on Communication Systems and Networks, IASTED*, 280–284. Marbella, Spain.

List of spacecrafts deployed from the International Space Station, 2020. https://en.wikipedia.org/wiki/List_of_spacecraft_deployed_from_the_International_Space_Station

Losik, L. (1995). Final report for a low-cost autonomous, unmanned ground station operations concept and network design for EUVE and other NASA Earth orbiting satellites. *Technology Innovation Series*, Publication 666: Center for EUVE Astrophysics, University of California, Berkeley, California, July.

Maini, K.A. and Agrawal, V. (2011). *Satellite Technology*, 2e. West Sussex: Wiley.

Maral, G. and Bousquet, M. (2005). *Satellite communications systems*, 4e. West Sussex: Wiley.

MOST. (2019). A tiny satellite probes the mysteries of universe. https://www.asc-csa.gc.ca/eng/stellites/most/Default.asp.

MOST inside. (2019). MOST Inside (2019[SC1]). https://images.squarespace-cdn.com/content/v1/5115d365e4b0b8b2ffe27887/1399032938148-KJ7EDZKYNIN51DITJJQ9/ke17ZwdGBToddI8pDm48kJ3JQ3iwwc4BFbzDBp0TFgd7gQa3H78H3Y0txjaiv_0fDoOvxcdMmMKkDsyUqMSsMWxHk725yiiHCCLfrh8O1z5QHyNOqBUUEtDDsRWrJLTm9NV9qK7dJxdxXunNrepQQJ993frrKLqPB2KgwTjQ701ds2_F8iBDySdn6PX6ZVQV/image-asset.gif.

Nikolova, K.N. (2002). *Antenna Noise Temperature and System Signal-to-Noise Ratio*. Hamilton, Ontario, Canada: McMaster University.

Northern Lights Software Associates, (2003), www.nlsa.com.

Pultarova, T. and Henry, C. (2017) OneWeb weighing 2,000 more satellites. *SpaceNews* (Feb. 24). https://spacenews.com/oneweb-weighing-2000-more-satellites/.

Reisenfeld, S., Aboutanios, E., Willey, K., Eckert, M., Clout, R. and Thoms, A. (2007) The Design of the FedSat Ka Fast-Tracking Earth Sydney: Cooperative Research Center of Satellite Systems, Faculty of Engineering, University of Technology.

Richharia, M. (1999). *Satellite Communications Systems*. New York: McGraw Hill.

Roddy, D. (2006). *Satellite Communications*, 4e. New York: McGraw Hill.

Saunders, R.S. (1993). *Antennas and Propagation for Wireless Communication System*. Sussex: Wiley.

Sentinel-6 Michael Freilich, (2021), https://www.jpl.nasa.gov/missions/sentinel-6.

Sheetz, M. (2019). Data from Investing in Space: Amazon Wants to Launch Thousands of Satellites So It Can Offer Broadband Internet from Space. CNBC (April 4). https://www.cnbc.com/2019/04/04/amazon-project-kuiper-broadband-internet-small-satellite-network.html (Accessed January 17, 2021).

Sklar, B. (2005). *Digital Communication*, 2e. New Jersey: Prentice Hall PTR.

The real benefit of SpaceX-Starlink highspeed internet (2022) https://cerexio.com/blog/the-real-benefit-of-spacex-starlink-high-speed-internet

Van Allen radiation belt (2020) https://en.wikipedia.org/wiki/Van_Allen_radiation_belt.

Walker, G., Matthews, J., Kuschnig, R. et al. (2003). The MOST asteroseismology mission: Ultra, precise photometry from space. *Publication of the Astronomical Society of the Pacific, USA* 115 (811): 1023–1035.

Zee, E.R. and Stibrany, P. (2002). The MOST microsatellite: a low-cost enabling technology for future space science and technology missions. *Canadian Aeronautics and Space Journal* 48 (1), Canada: 43–51.

2

Rain Attenuation

2.1 Rain Attenuation Concepts

Any signal of an appropriate frequency traveling between a ground station and a satellite, downlink or uplink, has to pass through Earth's atmosphere, which introduces certain impairments, resulting in signal power loss at any point in atmosphere of communication link, compared with the initial power (source power).

In satellite communication systems, the dominant propagation loss component is the free space loss, caused simply because of the distance between low Earth orbiting (LEO) satellite and the appropriate ground station. Furthermore, atmospheric losses are categorized into three components: ionospheric, tropospheric, and local. Since LEO satellites usually operate at S band (2.110–2.300 GHz) according to the acts the World Administrative Radio Conference, Malaga–Torremolinos, 1992, further only tropospheric effects will be treated, because the ionospheric effects are negligible at 2 GHz band, and the local effects are more related to Earth mobile systems. Thus, for practical purposes and performance evaluation, for LEO ground stations it is sufficient that only tropospheric impact be considered, seen from the LEO ground station perspective for communication in S-band.

The lower layer of the Earth's atmosphere, somewhere of around 10 km above the Earth's surface, is considered as troposphere. Most of the mass of the atmosphere is in the troposphere, as the densest region of atmosphere, but also it is the wettest region of the atmosphere, so, consequently all-weather changes on the Earth stem from this region.

The troposphere consists of a mixture of particles of different size and characteristics. Radio waves passing through such a medium will interact with this structure, resulting in a loss of radio wave power. This loss is known as *attenuation*, which stems from two factors – absorption and scattering – and depends on frequency. *Absorption* is the result of conversion from radio frequency energy to thermal energy because of the contact of waves with attenuating particles and *scattering* results from a redirection of the radio waves into various directions similarly because of the contact of radio waves with attenuating particles.

Scattering loss strongly depends on frequency and is only significant for operating frequencies above 10 GHz (Saunders, 1993) so it isn't an issue for the frequency band of 2 GHz related to LEO ground stations. Thus, only the absorption attenuation component of LEO ground stations will be considered here. The absorption attenuation component at any frequency primarily depends on a medium temperature and the humidity of particles structure.

Ground Station Design and Analysis for LEO Satellites: Analytical, Experimental and Simulation Approach,
First Edition. Shkelzen Cakaj.
© 2023 The Institute of Electrical and Electronics Engineers, Inc. Published 2023 by John Wiley & Sons, Inc.

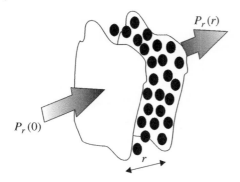

Figure 2.1 Rain attenuation.

Hydrometeors are liquid water particles, sometimes referring to as condensed water vapors in the atmosphere. Thus, rain, snow, hail, fog, or clouds are all examples of hydrometeors. All types of hydrometeors interfere with signal transmission. However, raindrops produce by far the highest attenuation because rain produces the highest humidity. More specifically, rain during higher average temperature (summer rain) manifests in the highest attenuation and is considered as the worst propagation case. This effect should be considered under the link budget estimation for the LEO ground station design.

Rain attenuation, also called rain fade, is illustrated in Figure 2.1, where $P_r(0)$ represents the signal power just in front of the rain region and $P_r(r)$ represents the signal power just at the leaving border line of the rain region of the distance r, which is the length of the region through the rain, know also the *rain path length* (Saunders, 1993).

Denoting the rain attenuation as A_R and expressing in decibels, follows:

$$A_R(dB) = 10 \log \frac{P_r(0)}{P_r(r)} \tag{2.1}$$

The most convenient way to express the attenuation caused by the any medium is applying the concept of specific attenuation, which is a very usual approach for cooper or fiber cables. Specific attenuation characterizes any medium and is expressed in [dB/km].

Rain-specific attenuation, denoted as γ, expressed in [dB/km], is a function of drops' radius r_d, drop size distribution $n(r)$, rain refractive index m (these three factors define rain structure), and frequency f, expressed as:

$$\gamma = F[r_d, n(r), m, f] \tag{2.2}$$

Different mathematical models (Lin, 1979; Crane, 1980) or latter (Dhaval et al., 2014) have been developed that confirm that the rain attenuation increases with rise in frequency and the rainfall rate (R), which is the rainwater that would accumulate in rain gauge situated at the ground in the region of interest (the location of the satellite ground station) (Pino et al. 2006). Rainfall rate is associated with a given percentage in time, which indicates the annual percentage p (%) during which a given value of rainfall rate R_p (mm/h) is exceeded. In most of Europe (for $p = 0.01\%$, corresponding to 53 minutes per year the given rainfall rate to be exceeded) a rainfall rate of 30 mm/h is suitable, except for some Mediterranean regions, where rainfall rates up to 50 mm/h have to be considered. Therefore, further analyses correspond to percentage of $p = 0.01\%$ for above values of rainfall rate.

Based on empirical models, it is found that γ (*specific rain attenuation*) depends only on R, where the rainfall rate R is measured on the ground in millimeters per hour (Saunders, 1993). From these empirical models, and approved by (ITU-R, 1992) (ITU 838, ITU-R P838-1, and Question 201/3), the usual form of expressing γ is:

$$\gamma = aR^b \tag{2.3}$$

where a and b are constants that depend on frequency, polarization, and average rain temperature. Table 2.1 shows values of a and b at various frequencies at 20° for both polarizations (ITU 838, ITU-R P838-1, and Question 201/3).

Table 2.1 Parameters of empirical rain attenuation model.

Frequency (GHz)	a_h	b_h	a_v	b_v
1.0	0.0000259	0.9691	0.0000308	0.8592
1.5	0.0000443	1.0185	0.0000574	0.8957
2.0	0.0000847	1.0664	0.0000998	0.9490
2.5	0.0001321	1.1209	0.0001464	1.0085
3.0	0.0001390	1.2322	0.0001942	1.0688
3.5	0.0001155	1.4189	0.0002346	1.1387
4.0	0.0001071	1.6009	0.0002461	1.2476

Table 2.2 Specific rain attenuation at 2 GHz (horizontal polarization) for different rainfall rates.

R [mm/h]	30	35	40	45	50
γ [dB/km]	0.00318	0.00375	0.00433	0.00491	0.00549

For illustration, the calculated specific rain attenuation values for 2 GHz band, under horizontal polarization, based on Eq. 2.3 and parameters from Table 2.1, are given in Table 2.2 (Cakaj and Malaric, 2006a).

Using the procedure applied above yields the specific rain attenuation for 2 GHz band under different rainfall rates R. To draw conclusions about the rain attenuation expressed in decibels, the next step is to find out the rain path length, expressed in km, since the specific rain attenuation is given in dB/km.

For the coherence purposes, the concept provided by (Saunders, 1993) is again applied and presented in Figure 2.2, providing the rain path geometry between the ground station and the satellite. Obviously, rain attenuation A_R (dB) depends on rain slant path length l_r. All heights in Figure 2.2 are considered above mean sea level. The effective rain height h_r in Figure 2.2 is the same as the height of the melting layer, where the temperature is $0°$ and h_s is the altitude of the satellite ground station. The representative values for effective rain height vary according to the latitude ϕ of the ground station location, given in kilometers (ITU, 618) by Eqns. (2.4) and (2.5). Since Europe belongs to the Northern Hemisphere, these values, expressed in km, are given by (Saunders, 1993):

$$h_r = 5 - 0.075(\phi - 23), \text{ for } \phi > 23 \tag{2.4}$$

$$h_r = 5, \text{ for } 0 \le \phi \le 23 \tag{2.5}$$

The rain path length from Figure 2.2 can be expressed as:

$$l_r = \frac{h_r - h_s}{\sin \varepsilon_0} \tag{2.6}$$

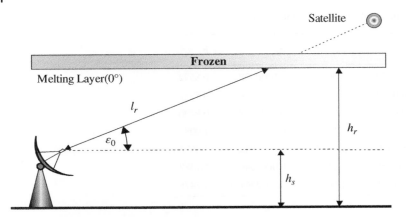

Figure 2.2 Rain path geometry.

h_s is the altitude of the ground station location. Finally, the rain attenuation A_R (dB) for rain path length l_r is:

$$A_R = \gamma l_r = aR^b l_r = aR^b \frac{\Delta h}{\sin \varepsilon_0} \tag{2.7}$$

and $\Delta h = h_r - h_s$. For paths in which the elevation angle is $\varepsilon_0 < 5^0$ it is necessary to account for the variation of the rain in the horizontal direction. This effect can be treated by implementing a reduction factor s, and then the attenuation is given by:

$$A_R = aR^b s l_r \tag{2.8}$$

According to [ITU, 618], this reduction factor of rainy path length is empirically defined as:

$$s = \frac{1}{1 + \dfrac{l_r \sin \varepsilon_0}{35e^{-0.015R}}} \tag{2.9}$$

where $e = 2.718$ is Neper constant. This equation is: valid for rainfall that does not exceed 0.01% of the time in an average year (around 53 minutes). Reduction factor s decreases with rainfall rate. Rain path length reduction factor is applicable only for elevation angle $\varepsilon_0 < 5^0$ (Saunders, 1993).

There are different models developed for rain attenuation calculations, but this one in my view is the simplest approach and provides very good inputs for link budget calculations and for further satellite ground station design. It follows the calculations procedure for typical LEO ground station.

2.2 Rain Attenuation for LEO Satellite Ground Station

The previous section described the mathematical approach related to the rain attenuation referred to the satellite ground station. Here we look at the appropriate calculations for the LEO station of a known location, and operating in the S-band.

Idea: The project MOST consisted of a LEO satellite and three ground stations, one of them in Vienna. The Vienna ground station system was set up at the Institute for Astronomy of the University of Vienna in cooperation with the Institute of Communications and Radio-Frequency

Engineering of the Vienna University of Technology (Keim et al., 2004) described previously under section 1.8, as LEO MOST and the ground station, given in Figures 1.23 and 1.25.

During the link budget calculations atmospheric attenuation had to be considered. The satellite link operated in 2 GHz band. In this band, the ionospheric effects are negligible, consequently, only the tropospheric effects are considered, more accurate, only rain absorption component, since also scattering is negligible at this band. The idea is to provide the mathematical calculation of the rain attenuation at 2 GHz, respectively at S-band, due to the communications of the Vienna satellite ground station with LEO satellite MOST (Cakaj and Malaric, 2006a).

Method: The (Saunders, 1993) approach is further applied. Rain attenuation depends on specific rain attenuation and the rain path length. Specific rain attenuation depends on the rainfall rate at the appropriate location, respectively depends on geographical position of the location where the station is installed. The rain path length strongly depends on the elevation angle under which the satellite is seen from the appropriate ground station. For calculation purposes, the ITU MODEL (ITU 838, ITU-R P838-1, Question 201/3) is applied.

The devices of the Vienna satellite ground station were installed at the Institute of Astronomy in Vienna. The downlink (from MOST satellite to the Vienna ground station) frequency is $f = 2232$ MHz (S-band). Austria belongs to the Northern Hemisphere and Vienna is at latitude $\phi = 48.2082^0$ and the altitude of the Institute where devices are installed is 203 m ($h_S = 203$ m http://www.stadtklima.de/webklima/CITIES/Europe/at/vienna/vienna.htm).

Substituting this latitude at Eq. (2.4) it is determined that $h_r = 3.125$ km (Cakaj and Malaric, 2007a). Then Δh is:

$$\Delta h = h_r - h_S = 2.922 \, [\text{km}] \tag{2.10}$$

The rain path length is:

$$l_r = \frac{2.922}{\sin \varepsilon_0} \, [\text{km}] \tag{2.11}$$

For an already-defined ground station site, the Δh can be written as:

$$\Delta h = l_r \cdot \sin \varepsilon_0 = const. \tag{2.12}$$

Finally, the known specific rain attenuation γ related to the area where the ground station is located, and the known rain path length l_r as a function of elevation angel ε_0 enable us to express the rain attenuation $A_R[\text{dB}]$ through elevation angle ε_0 for different rainfall rates R.

Results: Applying the given (Saunders, 1993) model for the data related to the Vienna ground station due to the communication with MOST satellite, the calculated rain attenuation results are given in Table 2.3 for different rainfall rates [R (mm/h)] and graphically presented in Figure 2.3 (Cakaj and Malaric, 2006a).

Rainfall rates are to be exceeded for 0.01% of annual time. Specific rain attenuation is considered for 2 GHz band, corresponding to Table 2.2. A rain path length reduction factors is applied for elevation angle $\varepsilon_0 < 5^0$, as given in Table 2.4.

Conclusion: For the band of 2 GHz, we found out that for the satellite ground station in Vienna dedicated for communications with LEO satellite, it is sufficient to consider the rain attenuation of 1 dB. This is applied within link budget calculations and the accuracy is confirmed through the proper work of the respective ground station in Vienna, where I operated the ground station for six months.

Table 2.3 Rain attenuation for different elevation angles.

Rainfall rate	R = 30 [mm/h]	R = 35 [mm/h]	R = 40 [mm/h]	R = 45 [mm/h]	R = 50 [mm/h]
Elevation	A_R[dB]	A_R [dB]	A_R [dB]	A_R [dB]	A_R [dB]
1°	0.536	0.615	0.690	0.765	0.834
2°	0.268	0.307	0.345	0.382	0.417
3°	0.179	0.205	0.230	0.255	0.278
4°	0.134	0.154	0.173	0.191	0.209
7°	0.087	0.100	0.114	0.127	0.141
10°	0.061	0.071	0.080	0.089	0.099
20°	0.031	0.036	0.041	0.045	0.050
30°	0.021	0.025	0.028	0.031	0.034
45°	0.015	0.017	0.020	0.022	0.024
60°	0.012	0.014	0.016	0.018	0.020
75°	0.011	0.013	0.014	0.016	0.018
90°	0.011	0.012	0.014	0.016	0.017

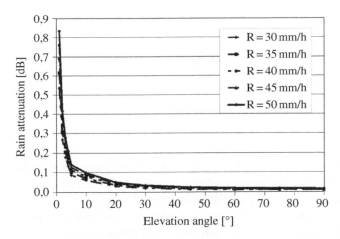

Figure 2.3 Rain attenuation of Vienna ground station for different elevation angles.

Table 2.4 Rainfall rate and rainy path horizontal length reduction factor.

R[mm/h]	30	35	40	45	50
s	0.894	0.888	0.883	0.875	0.870

2.3 Rain Attenuation Modeling for LEO Satellite Ground Station

Idea: Under the previous section we considered the rain attenuation for the link budget purposed for the satellite ground station in Vienna, dedicated for the MOST satellite. The idea behind this section is to analyze what happens from the rain attenuation point of view, if for the same purposes, the satellite is implemented somewhere else in Europe! For this generalization, the problem was modeled and applied for rain attenuation calculations for some of the capital cities in Europe, randomly chosen.

Method: Modeling methodology is applied, and based on the appropriate modeling flowchart a rain attenuation calculator is constructed. Rain attenuation depends on specific rain attenuation and the rain path length. Specific rain attenuation depends on the rainfall rate at the appropriate location – specifically, on geographical position of the location where the station is installed. The rain path length strongly depends on the elevation angle under which the satellite is seen from the appropriate ground station location. The rain path geometry from Figure 2.2 can be used to build the flowchart for general calculation of the rain attenuation for the different sites.

The modeling flowchart for general rain attenuation calculations is presented in Figure 2.2 (Cakaj and Malaric, 2007b) The acronyms in this figure stand for these input parameters: f– frequency, $a(f)$, $b(f)$- parameters of empirical rain attenuation model (Table 2.1), R – rainfall rate, ϕ – latitude of the ground station, h_S – altitude of the ground station, and ε_0 – is elevation angle. Further down the flowchart, the parameters are: γ – specific rain attenuation, h_r – the effective rain height, l_r – rain path length, s – rain path horizontal reduction factor, and finally A_R – rain attenuation. The appropriate flowchart is used for the design of the rain attenuation calculator given in Figure 2.4 (Cakaj and Malaric, 2007b). The rain attenuation calculator enables calculations of rain attenuation for different rainfall rates at different locations Figure 2.5.

For analytical purposes, some European cities are hypothetically chosen and supposed to be implementing a LEO satellite ground station. Further for the rain attenuation calculation, geographical data about these locations are needed. From http://earth.google.com we can find latitude and altitude of these cities (see Table 2.5). Data from Table 2.5 are also used for rain path length calculations.

Results: From Figure 2.2 it is obvious that the best propagation case happens under elevation angle of 90° since under

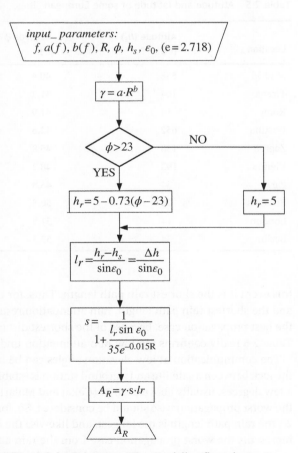

Figure 2.4 Rain attenuation modeling flow chart.

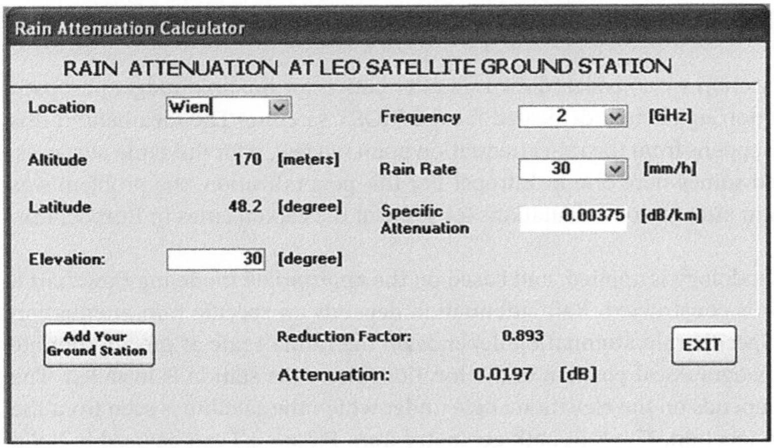

Figure 2.5 Rain attenuation calculator.

Table 2.5 Altitude and latitude of some European cities.

Location	Altitude (h_S) [m]	Latitude ϕ [°]	h_r [km]	$\Delta h = h_r - h_S$ [km]
Madrid	588	40.4	3.0.695	3.107
Tirana	104	41.3	3.625	3.521
Rome	14	41.9	3.582	3.568
Pristina	652	42.6	3.525	2.873
Zagreb	130	45.8	3.290	3.160
Vienna	193	48.2	3.113	2.920
Paris	34	48.8	3.060	3.026
Brussels	76	50.8	2.915	2.839
London	14	51.5	2.862	2.848
Berlin	34	52.5	2.786	2.752

this event it is the shortest rain path length. Thus, for the rainfall rate in Europe of $R = 30$ mm/h, and the shortest rain path length, rain attenuation results are presented in Table 2.6, representing the best propagation case, because of the shortest distance and the lowest rainfall rate considered. Table 2.6 really confirms too-low rain attenuation under these conditions.

The communication at low elevation angles can be hindered through natural barriers. Usually, the lock between a satellite and a ground station is established and lost in average at elevation above a few degrees, usually above 2° to 4° (Cakaj and Malaric, 2007b). For the link budget calculations, the worst propagation case should be considered. So, for communication under elevation angles of 2°, the rain path length is the longest, and likewise the highest rainfall rate for Europe of 50 mm/h represents the worst propagation case from the rain attenuation point of view. The results of the worst propagation case ($R = 50$ mm/h and $\varepsilon_0 = 2°$) in Table 2.7 are presented (Cakaj and Malaric, 2007a; Cakaj and Malaric, 2007b).

Table 2.6 Rain attenuation for $R = 30$ mm/h and $\varepsilon_0 = 90°$.

Location	Attenuation [dB]			
	2 [GHz]	2.5 [GHz]	3 [GHz]	3.5 [GHz]
Madrid	0.0088	0.0152	0.0253	0.0397
Tirana	0.0098	0.0184	0.0283	0.0444
Rome	0.0099	0.0186	0.0286	0.0449
Pristina	0.0082	0.0154	0.0236	0.0371
Zagreb	0.0089	0.0167	0.0257	0.0403
Vienna	0.0083	0.0156	0.0240	0.0376
Paris	0.0086	0.0161	0.0248	0.0388
Brussels	0.0081	0.0152	0.0234	0.0367
London	0.0081	0.0153	0.0235	0.0368
Berlin	0.0079	0.0148	0.0247	0.0357

Table 2.7 Rain attenuation under the worst propagation case $R = 50$mm/h and $\varepsilon_0 = 2°$.

Location	Attenuation [dB]			
	2 [GHz]	2.5 [GHz]	3 [GHz]	3.5 [GHz]
Madrid	0.4217	0.8141	1.3239	2.2337
Tirana	0.4694	0.9061	1.4735	2.5417
Rome	0.4747	0.9163	1.4902	2.5704
Pristina	0.3941	0.7606	1.2370	2.1337
Zagreb	0.4379	0.8260	1.3434	2.3172
Vienna	0.4000	0.7800	1.2535	2.1622
Paris	0.4122	0.7957	1.2941	2.2321
Brussels	0.3960	0.7582	1.2242	2.1117
London	0.3911	0.7550	1.2278	2.1179
Berlin	0.3795	0.7326	1.1914	2.0551

The results of the rain attenuation for frequencies of $f = 2$GHz and $f = 3.5$ GHz under $R = 50$ mm/h, are also graphically given in Figures 2.6 and 2.7.

Conclusion: This modeling approach confirms that the rain attenuation depends on geographical location where the satellite ground station is implemented, even for the same communication frequencies. For the worst propagation case, considering frequency band 2–2.5 GHz for different cities in Europe, the rain attenuation ranges from 0.37–0.91 dB, thus, remaining under 1 dB, the result of which stemmed from Vienna ground station analysis. Thus, for Central Europe, for S-band 2–2.5 GHz, it is proved that rain attenuation of 1 dB ensures reliable communication with sufficient margin. From the rain attenuation point of view, for the considered cities, the most convenient city is Berlin and the least convenient is Rome, simply because Rome is the lowest above the sea level.

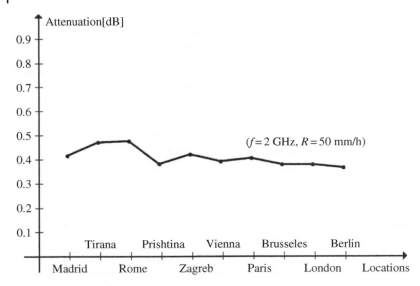

Figure 2.6 Rain attenuation for *f* = 2 GHz and *R* = 50 mm/h.

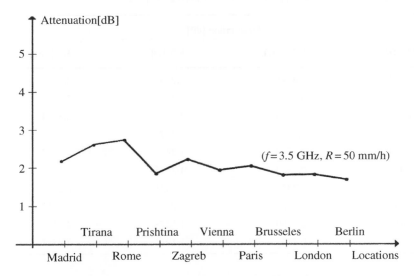

Figure 2.7 Rain attenuation for *f* = 3.5 GHz and *R* = 50 mm/h.

References

Cakaj, S. and Malaric, K. (2006a). *Rain Attenuation at Low Earth Orbiting Satellite Ground Station*, 247–250. Zadar, Croatia: *IEEE, Proc. 48th International Symposium ELMAR 2006 focused on Multimedia Systems and Applications*.

Cakaj, S. and Malaric, K. (2007a). Rigorous analysis on performance of LEO satellite ground station in urban environment. *International Journal of Satellite Communications and Networking* 25 (6): 619–643, Surrey, United Kingdom.

Cakaj, S. and Malaric, K. (2007c). *Rain Attenuation at Low Earth Orbiting Satellite Ground Station at S-band in Europe*, 17–20. Montreal, Canada: 18th IASTED International Conference on Modeling and Simulation.

Crane, R.K. (1980). Prediction of attenuation by rain. *IEEE, Transaction Communication Technology* COM-28 (September): 1717–1733.

Dhaval, P.P., Mitul, M.P., and Devendra, R.P. (2014). Implementation of ARIMA model to predict rain attenuation for KU-band 12Ghz frequency. *Journal of Electronics and Communication Engineering (IOSR-JECE)* 9 (1): 83–87, India.

ITU-R, (1992). Specific attenuation model for rain for use in prediction methods. Recommendation ITU-R P.838–1, Geneva.

Keim, W., Kudielka, V., and Scholtz, L.A. (2004). *A Scientific Satellite Ground Station for an Urban Environment*, 280–284. Marbella, Spain: International Conference on Communication Systems and Networks, IASTED.

Lin, S.H. (1979). Empirical rain attenuation model for earth satellite path. *IEEE Transaction Communication Technology* 27 (5): 812–817. https://doi.org/10.1109/TCOM.1979.1094458.

Pino, G.P., Riera, M.J., and Benarroch, A. (2006). Slant path attenuation measurement at 50GHz in Spain. *IEEE Antennas and wireless propagation letter* 4: 162–164.

Saunders, R.S. (1993). *Antennas and Propagation for Wireless Communication System*. Sussex: Willey.

3

Downlink Performance

3.1 Downlink Performance Definition

For the satellite communication systems, the performance of the receiving system, known as the downlink performance – the subject of the further elaboration – is commonly defined through a receiving system *Figure of Merit* as G/T_S (Sklar 2005):

$$T_S = T_A + T_{comp} \tag{3.1}$$

Here, G is receiving antenna gain, T_S is receiving system noise temperature, T_A is antenna noise temperature, and T_{comp} is composite noise temperature of the receiving system, including lines and equipment. The composite temperature depends exclusively on parameters of technical equipment and of interconnection lines characteristics. Otherwise, the antenna temperature T_A depends on external environment factors, including the sky background represented by its sky noise temperature denoted as T_C.

Obviously, for a given downlink antenna of the determined gain G, to determine the downlink performance, two factors of the denominator of G/T_S, must be calculated: composite noise temperature and antenna temperature. These are discussed in the next sections.

3.2 Composite Noise Temperature at LEO Satellite Ground Station

Idea: Noise introduces a fundamental limit on the performance of any communication system. Satellite communication systems are particularly susceptible to noise because of their inherent low received power, more concretely too low receive signal power at the LEO receiving ground station. During the link budget calculations, noisiness of the equipment and line loss must be considered. An alternative but equivalent way of expressing the noisiness of the equipment and line loss is through noise temperature. Composite noise temperature represents noise generated in all the components of the satellite receiving system chain. The idea is to calculate the composite noise temperature for a specific case, for known line and equipment parameters of the LEO satellite ground station.

Method: Here we apply the methodology for composite noise temperature calculations previously discussed under section 1.6. To illustrate these calculations, we use data from the Vienna ground station within MOST (Microvariability and Oscillation of Stars) project. The satellite link operates on 2GHz band (Cakaj and Malaric 2006b).

Ground Station Design and Analysis for LEO Satellites: Analytical, Experimental and Simulation Approach, First Edition. Shkelzen Cakaj.
© 2023 The Institute of Electrical and Electronics Engineers, Inc. Published 2023 by John Wiley & Sons, Inc.

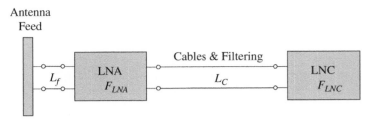

Figure 3.1 Simplified satellite receiving system.

Generally, the chain of devices of the LEO ground station receiving system does not stop at the low noise amplifier (LNA); it continuous further through cables and eventually filters toward the LNC [low noise (frequency) converter] and demodulator, as it is presented in the Figure 3.1.

In the general scheme of the satellite receiving system, the demodulator is behind the downconverter, but we will not consider that here, because its impact is of minor significance, based on Eq. (1.49). In the equation's denominator, the gains are multiplied, decreasing the impact of the further summary components. Applying parameters from the Figure 3.1 at Eq. (1.49) yields out:

$$T_{comp} = T_f + \frac{T_{LNA}}{G_f} + \frac{T_c}{G_f G_{LNA}} + \frac{T_{LNC}}{G_f G_{LNA} G_c} \tag{3.2}$$

where L_f is feed line loss and $G_f = 1/L_f$, L_C is loss caused by cables and filtering between LNA and LNC, then $G_C = 1/L_C$. Low noise amplifier's gain is G_{LNA}. Further, F_{LNA} and F_{LNC} are, respectively, noise figures for LNA and LNC. Since L_f, L_C, F_{LNA}, and F_{LNC} are usually known values (measured, calculated, or given by the manufacturer), then the respective temperatures will be calculated based on Eq. (1.39) and Eq. (1.42) as follows:

$$T_f = \left(L_f - 1\right) \cdot 290K \tag{3.3}$$

$$T_{LNA} = (F_{LNA} - 1) \cdot 290K \tag{3.4}$$

$$T_C = (L_C - 1) \cdot 290K \tag{3.5}$$

$$T_{LNC} = (F_{LNC} - 1) \cdot 290K \tag{3.6}$$

The downlink parameters for the equipment implemented at the LEO ground station in Vienna within MOST project are presented in Table 3.1 (Keim et al. 2004; Cakaj and Malaric 2006b).

Applying these parameters from the Table 3.1 at Eqs. (3.3)–(3.6), we get the noise temperatures for each component in Figure 3.1. These are presented in Table 3.2. (*Note:* Due to the calculation process, values expressed in decibels must be converted in linear values, to be substituted in Eqns. (3.3)–(3.6)). Finally, substituting data from Tables 3.1 and 3.2 at Eq. (3.2) we get the composite noise temperature for receiving satellite ground station.

Result: Applying methodology for the composite noise temperature calculations, we get the composite noise temperature for receiving system at the Vienna satellite ground station as (Cakaj and Malaric 2006b):

$$T_{comp} = 79.4 \, K \tag{3.7}$$

Conclusions: This composite temperature expresses the noisiness of the equipment and line loss of the receiving system of the ground station. This is one of the sum components of the system

Table 3.1 Downlink parameters of the Vienna satellite ground station.

Receiving frequency	2232 MHz	
Antenna gain		34.9 dBi
Feed line loss		0.4 dB
LNA noise figure		0.65 dB
LNA gain		41 dB
Loss (cables and filter.) Lc		4 dB
LNC noise figure		0.8 dB
LNC gain		32 dB
Data rate		38.4 Kbit/s
Receiver bandwidth	76 800 Hz	48.9 dBHz
Required S/N for 10E-5C BPSK		4.9 dB

Table 3.2 Noise temperatures for each system chain component.

T_f	29.78 K
T_{LNA}	46.82 K
T_c	438.45 K
T_{LNC}	58.65 K

temperature T_S. These calculations are a part of a link budget, especially for expressing the Figure of Merit of the receiving system. The methodology of calculations is applicable also for other satellite systems.

3.3 Antenna Noise Temperature at LEO Satellite Ground Station

Idea: The previous section considered the composite noise temperature. This section looks at antenna noise temperature and its impact on Figure of Merit. The idea is to compare the Figure of Merit under the case of medium attenuation presence, to the ideal case without medium attenuation, leaving the equipment structure completely unchanged.

Method: The comparison method and graphical interpretation are combined. Comparison means the analysis under different circumstances – in our case, between medium attenuation presence and no medium attenuation. The specific rain attenuation under different fall rate and frequencies, given in Figure 3.2, is used to calculate antenna temperature (Cakaj et al. 2011), which stems from Eq. (2.3) and Table 2.1. Since the goal of this section is just to make clear the impact of antenna temperature on Figure of Merit, we'll consider only the graphical interpretation.

The specific rain attenuation, marked with an arrow, refers to $R = 40$ mm/h and frequency of 2GHz, and it is $\gamma = 0.00407$ [dB/km] (Cakaj and Malaric 2006a).

Schematically, the satellite ground station receiving system and environment concept is presented in Figure 1.15 where T_C represents the sky noise temperature, T_m is medium temperature,

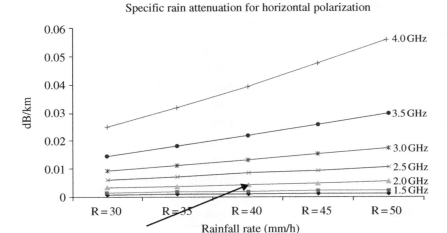

Figure 3.2 Specific rain attenuation.

and A is medium attenuation. The unwanted noise, injected via antenna is (kT_AB), and antenna temperature, is expressed by Eq. (1.29) (Saunders 1993).

LEO satellites move quickly over the ground station; thus, the ground station antenna must follow the satellite's movement, by changing its position in accordance with satellite's azimuth and elevation seen from the ground station. This means that the distance between the low Earth orbit (LEO) satellite and the appropriate ground station changes with elevation angle, and the same happens with antenna temperature. Antenna temperature for the ground station communicating with LEO satellite under the medium with specific attenuation of γ=0.00407[dB/km] is given in Figure 3.3 (Cakaj et al. 2011).

Figure 3.3 confirms the lowest antenna temperature under elevation of $\varepsilon_0 = 90^\circ$, since this is the shortest distance between the LEO satellite and the appropriate ground station.

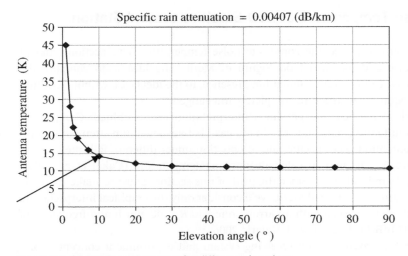

Figure 3.3 Antenna temperature for different elevations.

Let us consider two different cases of the satellite ground station where equipment and lines generate the composite noise temperature of $T_{comp} = 79.4\ K$ (see Eq. (3.7)). Next assume that the satellite is seen from the ground station under elevation of $\varepsilon_0 = 10°$, marked with an arrow in Figure 3.3.

The first case is supposed with no medium attenuation, $A = 0$, and from Eq. (1.29) we see that $T_A = T_c$, and for the highest sky temperature it is T_{A1}=10 K. The second case is under the medium attenuation given in Figure 3.3, where under $\varepsilon_0 = 10°$ antenna temperature is T_{A2}=15 K. Since the equipment are not changed, composite temperature remains as 79.4 K. From these conditions, we derive two system temperatures:

$$T_{S1} = T_{A1} + T_{comp} = 10 + 79.4 = 89.4K \tag{3.8}$$

$$T_{S2} = T_{A2} + T_{comp} = 15 + 79.4 = 94.4K \tag{3.9}$$

Results: Finally, we see that the antenna temperature depends on environmental circumstances, more accurately on medium attenuation included in between the link from/to the LEO satellite and the ground station. This impacts the Figure of Merit of the receiving system at the ground station. Considering the antenna gain of $G = 34.9$ dBi from Table 3.1, the variation on Figure of Merit is:

$$\left(\frac{G}{T_S}\right)_1 = 15.38 dB/K \tag{3.10}$$

$$\left(\frac{G}{T_S}\right)_2 = 15.15 dB/K \tag{3.11}$$

Conclusion: Antenna temperature depends on medium attenuation having direct impact on ground station performance, expressed through Figure of Merit. Presence of medium attenuation decreases the downlink performance.

3.4 Downlink Performance – Figure of Merit

Idea: From the previous sections, we have confirmed that the rain impacts on wave power propagated through the rain medium, and it is manifested as the signal power attenuation. But the rain also affects antenna temperature as one of the sum components of the system temperature, which strongly determines the performance of the ground station expressed through Figure of Merit (G/T_S).

Thus, the idea of this section is to draw conclusions about the impact of rain attenuation on the Figure of Merit (G/T_S), respectively, on the satellite ground station performance in the different European cities, randomly chosen. To exclude the impact of devices on such performance, it is assumed the hypothetical ground station, for all planned cities. So, the equipment parameters will be kept identical for all stations (T_{comp} is kept constant) to conclude exclusively about the rain impact on the Figure of Merit (Cakaj and Malaric 2008). Thus, the further step is for the hypothetical satellite ground station to be calculated the downlink performance for different cities in Europe, expressed by Figure of Merit (G/T_S) (Cakaj 2009).

Method: Since the further goal is to analyze and compare the effect of rain attenuation on the ground station performance, we assume that the same equipment and lines are implemented at all listed cities. This approach will eliminate equipment impact, so that the conclusions relate only to the rain attenuation impact on the downlink performance. Further, the Figure of Merit is calculated, and then the comparison methodology is applied toward further conclusions. The antenna temperature will be considered under the worst propagation case, respectively, under the elevation angle of $\varepsilon_0=2°$.

To calculate G/T_S, we should go further with system noise temperature T_S calculation, respectively, its sum components, antenna noise temperature T_A representing the external noise, and the composite noise temperature T_{comp} representing the internal noise, since it is:

$$\frac{G}{T_S} = \frac{G}{T_A + T_{comp}} \tag{3.12}$$

When an atmospheric absorptive process takes place (e.g., rain) the absorption process increases the antenna noise temperature. If the sky noise temperature is considered as T_C, the absorptive medium temperature as T_m, and the attenuation due the any absorptive process as A, then generally, the antenna noise temperature (downlink) impacted by the medium absorption is:

$$T_A = T_m\left(1 - 10^{-A/10}\right) + T_C 10^{-A/10} \tag{3.13}$$

Under the case of rain, it is $A = A_R$. (See: Table 2.7. Rain attenuation under the worst propagation case.) Generally, the range of needed parameters is given in Table 3.3 (Saunders 1993).

Thus, antenna temperatures for the worst propagation ($\varepsilon_0=2°$) case under the rain medium [Table 2.7, applied, $R = 50$ mm/h], and considering $T_m = 290K$ and $T_C = 10K$, is given in Table 3.4 (Cakaj et al. 2011).

Obviously, the antenna noise temperature differs for different cities at different operating frequencies.

Table 3.3 The range of parameters related to antenna noise temperature.

Rain medium temperature (T_m)	(275–290) K
Sky temperature (T_C)	(3–10) K
Rain attenuation (A_R)	(1–3) dB

Table 3.4 Antenna noise temperature.

	Antenna temperature [K]			
Location	2 [GHz]	2.5 [GHz]	3 [GHz]	3.5 [GHz]
Madrid	34.9	56.1	80.9	120.4
Tirana	37.6	60.8	87.7	129.6
Rome	37.9	61.8	88.4	130.6
Pristina	33.4	53.3	76.9	114.8
Zagreb	35.3	56.7	81.8	121.6
Vienna	33.7	53.9	77.7	115.9
Paris	34.4	55.2	79.6	118.5
Brussels	33.1	52.9	76.4	113.9
London	33.2	53.1	76.4	114.2
Berlin	32.5	51.9	74.7	111.8

The next step is the calculation of the composite temperature T_{comp}, which represents accumulated internal noise generated by the lines and equipment of the satellite ground station. For this purpose, the hypothetical ground station has the parameters given in Table 3.1. The calculation of T_{comp} is clarified under 3.2. Applying the parameters from Table 3.1 yields:

$$T_{comp} = 79.4\,K \tag{3.14}$$

Finally, applying composite temperature as $T_{comp} = 79.4\,K$ and antenna temperature T_A from Table 3.4 (which is related to the rainfall rate of 50 mm/h), then antenna gain of 40 dB at Eq. (3.12), stems out the Figure of Merit (G/T_S) expressing the impact of the rainfall rate on the antenna noise temperature and consequently on the LEO satellite ground station performance. The same also applies for the other rainfall rates of 30 mm/h and 40 mm/h.

Results: The complete results for different rainfall rates are given in Table 3.5 (Cakaj and Malaric 2008).

Calculations show that for the same frequency and the same rainfall rate, the difference in Figure of Merit among cities is less than 0.2dB.

$$\Delta(G/T_S) < 0.2 dB/K \tag{3.15}$$

This means that the ground station's performance within central Europe does not strongly depend on location. This fact creates opportunities for implementing the LEO ground stations in Europe for different purposes. Further, if we consider the same frequency band and different rainfall rates, then the differences on Figure of Merit for different cities from Table 3.5 are as follows (Cakaj 2009):

$$\Delta(G/T_S) < 0.5 dB/K \text{ for 2GHz} \tag{3.16}$$

$$\Delta(G/T_S) < 1 dB/K \text{ for 3GHz} \tag{3.17}$$

$$\Delta(G/T_S) < 1.6 dB/K \text{ for 4GHz} \tag{3.18}$$

Further for Pristina (capital of Kosovo), Figure of Merit on dependence of frequency is given in Figure 3.4. The main improvement on the Figure of Merit could be achieved by the receiving

Table 3.5 Figure of Merit for different European cities (G/T_S) (dB/K).

Frequency (GHz)	2			3			4		
Rain Rate	30 mm	40 mm	50 mm	30 mm	40 mm	50 mm	30 mm	40 mm	50 mm
Pristina	19.82	19.62	19.44	18.86	18.40	17.99	17.31	16.48	15.87
Rome	19.70	19.47	19.27	18.61	18.10	17.67	16.96	16.14	15.56
Vienna	19.81	19.61	19.43	18.85	18.38	17.97	17.29	16.46	15.85
Berlin	19.85	19.65	19.47	18.91	18.45	18.05	17.38	16.55	15.93
Brussels	19.83	19.63	19.45	18.88	18.41	18.00	17.33	16.50	15.89
London	19.83	19.63	19.45	18.88	18.41	18.00	17.33	16.50	15.88
Madrid	19.78	19.57	19.38	18.77	18.30	17.88	17.18	16.36	15.76
Paris	19.79	19.59	19.40	18.80	17.33	17.92	17.23	16.40	15.79
Tirana	19.70	19.48	19.28	18.63	17.12	17.70	17.98	16.16	15.56
Zagreb	19.77	19.56	19.37	18.76	17.27	17.86	17.16	16.33	15.73

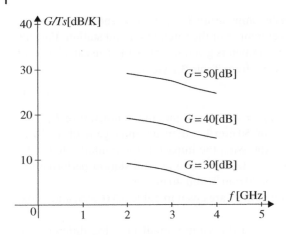

Figure 3.4 Figure of Merit for different receiving antenna gain (Pristina).

antenna gain. For the same parameters from Table 3.5, just by changing the receiving antenna gain for Pristina significantly improves the Figure of Merit on dependence of frequency.

Conclusion: The rain attenuation analyses are too important for link budget considerations due to the ground station design. For Central Europe at range of 2–4 GHz, the rain attenuation varies from 0.5 dB to 1.6 dB. The performance of the ground station within central Europe does not strongly depend on location. The difference on downlink Figure of Merit among all considered cities in central Europe for S-band in average is less than 0.2 dB under the same rainfall rate, or on average less than 1 dB for different rainfall rates. Rain attenuation of 1 dB remains sufficient to be considered within link budget calculations at S-band. Through the case of Pristina, it is confirmed that changing the receiving antenna gain achieves the most improvement on the Figure of Merit.

3.5 Downlink Performance: Signal-to-Noise Ratio (S/N)

Idea: There are two types of ground stations: single-antenna systems and double-antenna systems (Elbert 1999). At the single-antenna concept, the separation of the transmission and reception is achieved by means of duplexer. This section uses the link budget analysis to mathematically compare these two concepts to determine which one should be implemented. For analytical purposes, we consider data for the Vienna ground station, given in Table 3.1 (Keim et al. 2004; Cakaj and Malaric 2007a).

Method: Comparison method is applied. As parameters to be used for comparison are Figure of Merit (G/T_S), signal-to-noise ratio (S/N), and downlink margin (DM), which will be calculated for both cases, starting with double-antenna system and then implementing the duplexer for a single-antenna system. For both cases, parameters of components must be applied. The block diagram of the downlink segment at the Vienna LEO ground station within MOST space observation project is presented in Figure 3.5. The acronyms in Figure 3.5 are: BPSK (binary phase shift keying), LNA (low noise amplifier), LO (local oscillator), H (horizontal polarization), and V (Vertical polarization).

Applying parameters from Table 3.1, under section of composite noise temperature, the result is:

$$T_{comp} = 79.4K \tag{3.19}$$

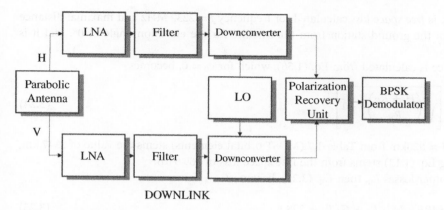

Figure 3.5 Block diagram of the downlink segment.

The antenna temperature depends on where the antenna is looking. The antenna itself is considered lossless. Assuming that the antenna sees the sky without medium attenuation, the solid angle subtended by the noise source (sky) is much larger than antenna beam angle, so, in this case, antenna noise temperature T_A is equal to the sky noise temperature T_C. This is *the best propagation* case, where $T_A=T_C=10$ K.

For the ground station, which is communicating on 2 GHz band as *the worst propagation* case, heavy rain of 50 mm/h is the most influent factor on antenna temperature. It is concluded that attenuation of $A = 1$ dB is sufficient to be considered (see section 2.1). Thus, considering frequency band, rain attenuation, and rain medium temperature of $T_m= 290$ K, the antenna temperature for *the worst propagation* case is calculated as $T_A= 65.5$ K. [*Comment:* Looking at Table 3.4, for the Vienna ground station, the antenna temperature is 33.7 K, but this relates to the specific rain attenuation of 0.4 dB/km from the Table 2.7.] Considering the above, calculated antenna temperature and composite temperature from Eq. 3.19, the system temperature and *Figure of Merit* are:

$$19.5dBK \leq T_S \leq 21.6dBK \tag{3.20}$$

$$13.3dB/K \leq G/T_S \leq 15.4dB/K \tag{3.21}$$

BPSK modulation for the communication between the ground station and the satellite is used. Satellite MOST transmits toward ground station with these parameters: Transmit power of 0.5 W (-3dBW), loss of 2 dB and, antenna gain of 0dBi (Zee and Stibrany 2002). In link budget calculations, of the greatest interest is receiving system signal-to-noise ratio $[(S/N)$ or $(S/N_0)]$ expressed by *range equation* as:

$$\frac{S}{N_0} = \frac{EIRP(G/T_S)}{kL_S} \tag{3.22}$$

where *EIRP* is effectively isotropic radiated power from the transmitter. Considering that $N=N_0B$, $N_0 = kT$ where, N_0 is spectral noise density, B receiver bandwidth, $k = 1.38 \cdot 10^{-23}$W/HzK is Boltzmann's constant, and expressing Eq. (3.22) in decibels yields:

$$\frac{S}{N_0}(dB) = EIRP - L_S + G/T_S + 228.6 \tag{3.23}$$

In Eq. (3.23), L_S is *free space loss* calculated for frequency $f = 2232$ MHz and maximal distance $d_{max} = 3347$ km of the ground station from the satellite for the elevation angle of 0°, and it is 169.9 dB.

Maximal distance is calculated from Eq. (1.56), which for $\varepsilon_0 = 0$, becomes:

$$d_{max} = R_E \left[\sqrt{\left(\frac{H + R_E}{R_E}\right)^2 - 1} \right] \tag{3.24}$$

where applying $H = 820$km from Table 1.7 (MOST orbital elements) stems the value of 3347 km, and then applying Eq. (1.13) stems from the free space loss of 169.9 dB.

If we consider other losses L_0, then Eq. (3.23), becomes:

$$\frac{S}{N_0}(dB) = EIRP - L_S - L_0 + G/T_S + 228.6 \tag{3.25}$$

For the Vienna ground station, since it was the low-cost project, considering all factors, 5 dB are added related to other losses (polarization loss, miss pointing, etc.). Finally, for the worst propagation case and the worst system temperature, considering the equipment's parameters from Table 3.1, Table 3.6 lists the downlink budget under assumption of double-antenna system implementation.

The downlink margin is defined as:

$$DM = (S/N)_r - (S/N)_{rdq} \tag{3.26}$$

where r, rqd mean received and required. Let me just add that Eq. (1.19) is applied for (S/N).

Further, considering the idea, we have to compare link budget under two scenarios, so the following relates to a single-antenna system, where we consider the same equipment and the same propagation condition, and only add the duplexer for uplink and downlink signal separation.

Table 3.6 Downlink budget for double antenna system.

Transmit power	−3	dB
Loss	−2	dB
MOST antenna gain	0	dBi
EIRP	−5.0	dBW
Total propagation loss	−174.9	dB
Received isotropic power	−179.9	dBW
Antenna gain	34.9	dBi
System noise temperature	21.6	dBK
Figure of Merit (G/T_S)	13.3	dB/K
S/N_0	62	dB
Receiver bandwidth	48.9	dBHz
S/N	13.1	dB
Required S/N (BPSK)	4.9	dB
Downlink margin	8.2	dB

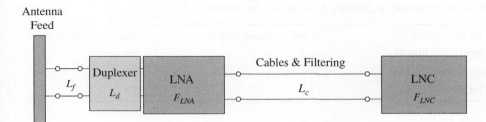

Figure 3.6 Simplified satellite receiving system with duplexer.

The chain of devices of the receiving system from Figure 3.1, in case the duplexer, is implemented in between the feeder and low noise amplifier, as depicted in Figure 3.6.

The composite noise temperature for receiving system with duplexer will be as:

$$T_{comp} = T_f + \frac{T_d}{G_f} + \frac{T_{LNA}}{G_f G_d} + \frac{T_C}{G_f G_d G_{LNA}} + \frac{T_{LNC}}{G_f G_d G_{LNA} G_c} \quad (3.27)$$

where T_d is duplexer's noise temperature as:

$$T_d = (L_d - 1) \cdot 290K \quad (3.28)$$

where $L_d (G_d = 1/L_d)$ is the duplexer's attenuation. Considering duplexer's attenuation $L_d = 3dB$ and applying the same methodology as for the case without duplexer, it is calculated the system temperature and *Figure of Merit* as:

$$26.5dBK \leq T_S \leq 27dBK \quad (3.29)$$

$$7.9dB/K \leq G/T_S \leq 8.4dB/K \quad (3.30)$$

The downlink budget for the single antenna system with duplexer follows in Table 3.7.

Results: Under method, we mentioned that three factors would be compared: Figure of Merit, S/N ratio, and downlink margin. From the above calculation, these parameters are given in Table 3.8.

Table 3.7 Downlink budget for single antenna system.

Transmit power	−3	dB
Loss	−2	dB
MOST antenna gain	0	dBi
EIRP	−5.0	dBW
Total propagation loss	−174.9	dB
Received isotropic power	−179.9	dBW
Antenna gain	34.9	dBi
System noise temperature	27	dBK
Figure of Merit (G/T_S)	7.9	dB/K
S/N_0	56.6	dB
Receiver bandwidth	48.9	dBHz
S/N	7.7	dB
Required S/N (BPSK)	4.9	dB
Downlink margin	2.8	dB

Table 3.8 Parameter's comparison: double vs. single antenna system.

Parameters for comparison	Double antenna system	Single antenna system
Figure of Merit (G/T_S) [dB/K]	13.3	7.9
Signal-to-noise ratio (S/N) [dB]	13.1	7.7
Downlink margin (DM) [dB]	8.2	2.8

Table 3.8 shows that the downlink margin is 8.2 dB and S/N =13.1dB without a duplexer. Then by implementing the duplexer, the downlink margin becomes 2.8 dB and S/N =7.7dB.

Conclusion: For the system planned for both transmit and receive operation and unattended work, we analyzed whether the single antenna or double antenna system with duplexer should be used. It is confirmed that the downlink margin and S/N ratio is much better with the double antenna concept than with duplexer implementation. There is a risk of downlink margin loss by duplexer implementation in the front end. This leads to the preference for the double antenna system. Thus, for uplink, another antenna should be used.

3.6 Downlink and Uplink Antenna Separation

Idea: Based on the previous section, it is better to implement a double-antenna system for an unattended ground station. But this conclusion begs the question, if two antennas must be implemented, for downlink and uplink, how far apart should antennas be so as to not disturb each other (Cakaj and Malaric 2007b)? The clarification follows.

Method: Measurement through experiment is applied. Uplink frequency is 2055 MHz and the downlink frequency is 2232 MHz. The isolation experiment setup is shown in Figure 3.7.

As the first step, each antenna should be tested separately and then installed in the tracking system (Rezaei and Hakkak 2005). At the ground station, the uplink uses 50 W power amplifier on FM modulation at frequency of 2055 MHz. Four Yagi antennas feed in phase are used for transmission.

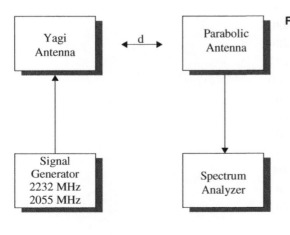

Figure 3.7 Isolation measurement setup.

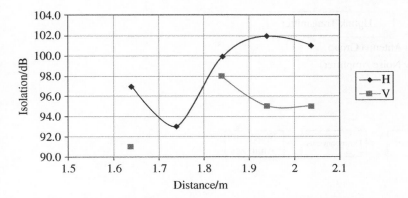

Figure 3.8 Isolation between uplink antenna group and downlink antenna.

Yagi antenna group was fixed beside the downlink 3 m parabolic antenna. By changing the distance, the power level is measured at the spectrum analyzer for adjusted power at a signal generator. This is done for both, transmit and receive frequencies (2055 MHz and 2232 MHz) in both polarizations.

Results: The measured isolation in between downlink and uplink antennas, by changing distance from 1.5 m up to 2.1 m, is presented in Figure 3.8 (Cakaj and Malaric 2007b).

Conclusion: It is confirmed that the double-antenna system has noticeably better downlink margin compared with the solution of a single parabolic antenna and a duplexer, but that approach should be followed by the above experiment to find out the physical distance between them. In this case, at the distance of 1.84 m, the isolation between downlink and uplink antennas is at least 98 dB. This methodology is applicable for other frequency bands.

3.7 Desensibilization by Uplink Signal at LEO Satellite Ground Station

Due to the full duplex operation, the uplink signal is permanently present while the downlink is receiving data from the satellite. Strong signals near the passband of an LNA can reduce the sensibility of the LNA, however, and consequently the sensibility of the entire receiving system.

Idea: Considering that the downlink could be desensibilized by the uplink, the idea is to experimentally check if the uplink disturbs the LNA, and consequently in fact disturbs the performance of the downlink receiving system. The simplest approach is to check the performance under two events: turning the uplink off and then turning it on.

Method: The experiment setup and methodology for checking an eventual desensibilization of the receiving system caused by the presence of the permanent uplink signal is presented in Figure 3.9 (Cakaj et al. 2007).

In Figure 3.9 the spectrum analyzer is used as power meter. The f_t is the uplink signal frequency. The desensibilization effect is to be measured at the IF output of the downconverter as the receive signal due to the Sun noise is applied, presented in Figure 3.10 as the upper trace (Cakaj et al. 2007). For LEO satellites, the ground station must track the satellite during its short flyover period when it is range, about 5–15 minutes at a time. A ground satellite station antenna must precisely follow the

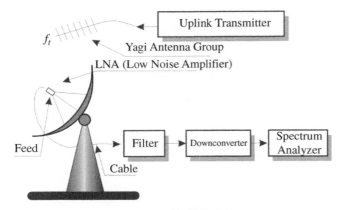

Figure 3.9 Desensibilization measurement setup.

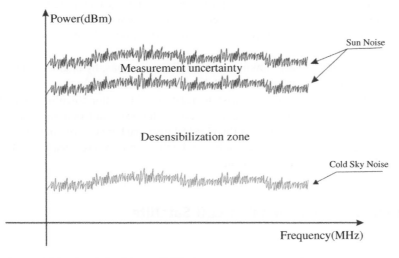

Figure 3.10 Concept of desensibilization measurement result.

quick-moving satellite with pointing accuracy of 0.5°. Mismatch in pointing will reduce the desired signal level strength. This mismatch in the antennas' pointing is also considered during this experiment and is represented in Figure 3.10 as the measurement uncertainty zone.

The methodology of experiment execution follows. The uplink signal is turned off. The spectrum analyzer records only downlink received noise signal. First, the antenna will be pointed to the cold sky, and the cold sky noise level will be recorded. Then, the antenna will be pointed to the Sun, while the upper power level will be recorded as the Sun noise level.

After these records, the uplink signal will be turned on and again will be measured receiving Sun noise signal. If under the permanent presence of the uplink signal the received Sun noise signal is within the measurement uncertainty zone, the LNA and consequently the receiving system is not desensibilized. If the received signal is under the lower level of uncertainty zone (desensibilization zone), the LNA and consequently the receiving system is desensibilized.

Results: The results of this experiment executed at the Vienna ground station are presented in Figure 3.11 (Cakaj et al. 2007). This enables the comparison with the idea presented in Figure 3.10.

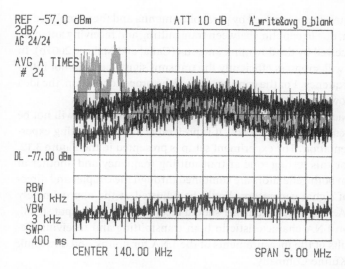

REF -57.0 dBm   ATT 10 dB   A_write&avg B_blank
2dB/
AG 24/24

AVG A TIMES
24

DL -77.00 dBm

RBW
10 kHz
VBW
3 kHz
SWP
400 ms

CENTER 140.00 MHz   SPAN 5.00 MHz

Figure 3.11 Desensibilization by uplink signal measurement result.

At satellite ground station, the uplink signal output power is 50 W. The bottom signal represents the received noise power density when the antenna is pointed to the cold sky. Then the antenna is pointed to the Sun. As already mentioned, there is a residual pointing error of the antenna.

Thus, the highest and the lowest Sun noise power output level at the downconverter's output of several pointing attempts are recorded as two traces (a little bit darker in figure). The difference of around 2 dB between these two traces represents the measurement uncertainly, including pointing error and instruments accuracy. In order to test the influence of the uplink signal on the sensibility of the downlink receiving system, the uplink transmitter is turned on. In this case, the downconverter's recorded output trace somehow falls in between two earlier recorded traces, and has two peaks.

On the left upper corner in Figure 3.11, there are two peaks and then the signal continually falls off at approximately the same level of the records done when the uplink was off (it is in fact in the middle). These peaks represent generated intermodulation products, which in the next section will be confirmed that do not fall in the receiver's passband and do not disturb the receiving system (Cakaj et al. 2005a). Since, the last recorded trace (when the uplink is turned on) is in between the two traces recorded when the uplink was off, we can conclude that the sensibility of the receiving system is not reduced by the uplink signal presence. In case of the desensibilization, the trace recorded when the uplink was turned on had to be under the lower trace recorded when the uplink was off.

Conclusion: Our experiment confirmed that the receiving system at the LEO ground station is not desensibilized by the permanent presence of uplink signal. This ensures a proper work of the ground station and normal data flow in both directions. This is an easy executable measurement to check if desensibilization is present and if it disturbs the receiving system.

3.8 Downlink and Uplink Frequency Isolation

In the previous experiment, coarse testing was only from a desensibilization aspect, based on Sun noise power, which provides a sufficient fact for the proper operation. However, from the frequency interference viewpoint, deeper treatment is required.

Idea: The weak signals from the satellite are received by parabolic antenna and then amplified by the LNA. In a double antenna system, due to coupling between transmitting and receiving antenna, the transmitted signal can also be received by the downlink antenna (Cakaj and Malaric 2007b). In this case, filters should be used that will suppress efficiently the transmit signal but will not introduce significant loss at the receive frequency. For the confirmation of the proper operation, the idea is to measure the isolation on the downlink at both frequencies, uplink and downlink.

Method: The experimental approach is applied. To ensure that the downlink signal will not be blocked by transmit signal (uplink) coupled into the LNA, it is preferable for the following experiment to be executed before implementation. The experiment setup is presented in the Figure 3.12.

Two signal generators are used for this testing, one at transmitting frequency and another at receiving frequency. Two signals from both generators are combined through the coupler and directed to the LNA. The output level of signal generators should be adjusted at the approximately expected level at the input of the LNA. A spectrum analyzer should record the LNA's output signal. The sensibility reduction depends on LNA characteristic in both transmitting and receiving frequencies. Figure 3.13 shows the results of the measurements of the LNA for uplink and downlink frequencies (KU 222) produced by Kuhne Company.

Results: Records of spectrum analyzer show blocking or isolation of transmit signal, given in Figure 3.13, where it is obvious there is at least the 30 dB isolation between uplink and downlink frequency (Cakaj, Keim and Malaric 2007).

Conclusion: The isolation of at least 30 dB ensures the proper functionality of the downlink.

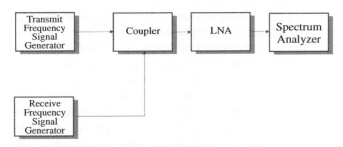

Figure 3.12 Blocking experiment setup.

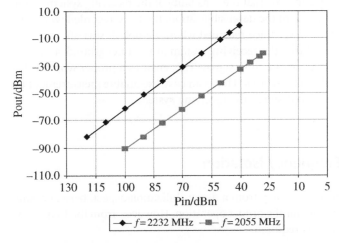

Figure 3.13 Performance measurement of low-noise amplifier.

3.9 Sun Noise Measurement at LEO Satellite Ground Station

Receiving LEO ground station systems must frequently process very weak signals. The noise added to the signals by the system components tends to obscure such signals. In the course of receiver evaluation, measurements of the noise contributions are necessary, which impact the system Figure of Merit, and consequently the downlink performance.

Idea: The idea here is to experimentally measure Figure of Merit of the LEO ground station and to compare it with the results from mathematical approach given under the link budget calculation under Sections 3.5 and 3.6. So, the main goal is to conclude whether the mathematical results agree with experimental result. Measurement experiment will be based on Sun radio flux density, further described (Cakaj et al. 2005b).

Method: Experimental approach is applied. The measurement is based on the *hot/cold method*, also known as the *Y-factor method*. The measurement setup is presented in Figure 3.14. Here, DUT means *device under test*.

The "Y-factor" noise figure measurement technique uses two noise sources at two different temperatures to determine the noise temperature T_{dut} of the DUT. With each noise source R_S connected to the DUT, the output powers corresponding to the different temperatures T_{hot} and T_{cold} are measured. The powers are termed P_{hot} and P_{cold}. The Y-factor is defined as the ratio of these powers:

$$Y = \frac{P_{hot}}{P_{cold}} \tag{3.31}$$

If we consider the DUT has gain $G = 1$ and bandwidth B, and further assume that there are no losses resulting from cables, no mismatch losses, then based on Eq. (1.30) P_{hot} and P_{cold} are:

$$P_{hot} = k(T_{hot} + T_{dut})B \tag{3.32}$$

$$P_{cold} = k(T_{cold} + T_{dut})B \tag{3.33}$$

Solving by T_{dut} Eq. (3.32) and Eq. (3.33) yields:

$$T_{dut} = \frac{T_{hot} - YT_{cold}}{Y - 1} \tag{3.34}$$

From Eq. (3.34) for known temperatures and measured Y, the noise temperature T_{dut} can be obtained.

For satellite communication systems, to obtain G/T_S one could determine G and T_S separately, which needs elaborated measurements. It is much easier to obtain the ratio (G/T_S) by a single measurement based on the solar radio flux density. The Sun is considered a *hot source* and the cold sky is a *cold source*. The principle of determination (G/T_S) is to measure the increase on noise power that occurs when the antenna is pointed first at a cold region of the sky (Temp. of around 3 K–10 K) and then moved to a strong source of known radio flux density, usually the Sun (temperature of around 13 000 K). Thus, the Y-factor is:

$$Y = \frac{P_{sun}}{P_{coldsky}} \tag{3.35}$$

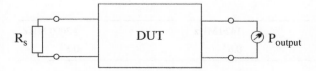

Figure 3.14 Noise measurement with Y-factor method.

The measured power consists of two components: the power generated by the receiving system itself (kT_sB) and the power coming from the external radio source. In case of the cold sky, because of the low temperature (average around 6 K), the solar flux density can be considered as very low or negligible. Then, the measured power $P_{coldsky}$ can be expressed as

$$P_{coldsky} = kT_sB \tag{3.36}$$

The power P_{sun} when the antenna is pointed to the Sun is (Flagg 2003):

$$P_{sun} = kT_sB + F_{sun}A_eBL \tag{3.37}$$

where ($F_{sun}A_eBL$) is the noise power resulting from solar radiation. In Eq. (3.37), F_{sun} is the solar radio flux density at the test frequency f and expressed in (W/m^2Hz), L is a beam size correction factor, B is the bandwidth of the system and A_e ($A_e = \lambda^2G/4\pi$) is the antenna's effective area. Since the solar radiation is randomly polarized; an antenna designed for single polarization will by high probability receive half of incoming power (Hoch 1979), as:

$$F_{sun.single.pol} = \frac{F_{sun}}{2} \tag{3.38}$$

Thus, for a single polarization measurement P_{sun} will be:

$$P_{sun} = kT_sB + \frac{F_{sun}}{2}A_eBL \tag{3.39}$$

Substituting Eq. (3.38) at Eq. (3.37), and further at Eq. (3.35), yields:

$$\frac{P_{sun}}{P_{coldsky}} = Y = 1 + \frac{F_{sun}A_eBL}{2kT_sB} \tag{3.40}$$

Considering $A_e = \lambda^2G/4\pi$ and substituting at Eq. (3.40), gives:

$$Y = 1 + \frac{F_{sun}\lambda^2GL}{8\pi kT_S} \tag{3.41}$$

Solving by (G/T_S), Eq. (3.42) the *Figure of Merit of the Receiving System* is:

$$\frac{G}{T_s} = \frac{8\pi k}{F_{sun}L\lambda^2}(Y-1) \tag{3.42}$$

For Eq. (3.42), the value of Y will be obtained by measurement. Further L and F_{sun} are to be investigated. The beam size correction factor L is expressed by:

$$L = 1 + 0.38\left(\frac{\theta_{sun}}{\theta}\right)^2 \tag{3.43}$$

and is dependent on antenna beamwidth. In Eq. (3.43), θ_{sun} is the diameter of the radio sun in degrees at frequency f and θ is the antenna 3 dB beamwidth at frequency f. The diameter of the radio Sun θ_{sun} is frequency dependent, as it is presented in the Table 3.9 (Cakaj et al. 2005b).

Table 3.9 Radio sun diameter table.

f	400 MHz	1420 MHz	$\geq$3000 MHz
θ_{sun}	0.7^0	0.6^0	0.5^0

The beamwidth θ of a parabolic antenna is: $\theta = 70\lambda/d$ where λ is the wavelength corresponding to frequency $f (c = f\lambda)$ and d is the diameter of the parabolic antenna. For antenna of diameter $d = 3$ m and the downlink frequency $f = 2232$ MHz, then, it is $\theta = 3.13°$. Further $0.38(\theta_{sun}/\theta)^2 \sim 0.01$, which leads to beam size correction factor $L = 1$ and Eq. (3.42). Then the *Figure of Merit* for the receiving system becomes:

$$\frac{G}{T_s} = \frac{8\pi k}{F_{sun}\lambda^2}(Y-1) \tag{3.44}$$

The next term needed is *solar radio flux density* (F_{sun}) at the test frequency. The USAF (United States Air Force) Space Command runs a worldwide solar monitoring network and measures the solar radio flux density. Since the actual solar radio flux density is always available, it may be used to make reasonably accurate measurements of receiving system performance.

The accuracy of the determination of (G/T_S) is dependent on the accurate measurement of Y (Cakaj et al., 2005b; Morgan, 2018). The easiest measurement technique is to use a power meter connected to the downconverter's IF output respectively receiver's input. Whatever technique is used, the Y-factor needs to be measured several times and an average must be taken, although Figure 3.15 makes clear that the solar radio flux trend is constant over time. Solar radio flux density must be determined at approximately the same time as the measurements of (G/T_S) are done (Cakaj et al., 2005b; Morgan 2018).

Solar radio flux density is available at any time. The USAF Space Command measures the solar radio flux density at eight "standard" frequencies (www.ips.gov.au): 245, 410, 610, 1415, 2695, 4995, 8800, and 15 400 MHz. These data are for Quiet Solar, and the values are expressed in *sfu (solar flux unit)* units (Monstein 2002). The solar flux density is also expressed in *Jansky unit*, where:

$$1 Jansky = 10^{-26} W/m^2 Hz$$
$$1 sfu = 10^4 Jansky = 10^{-22} W/m^2 Hz$$

A few records of the solar radio flux density expressed in *solar flux units* and recorded by dates are presented in Table 3.10.

These data are recorded by Australian Space and Weather Agency at Learmonth observatory. At the same web page [www.ips.gov.au] could also be found the solar flux interpolated values for the following frequencies: 1300, 1540, 1707, 2300, 2401, 2790, 5625, 6000, 8000, 8200, 9410, and 10 400 MHz. Table 3.11 presents solar radio flux interpolated values for respective frequencies for the same dates as from Table 3.10.

If the operating frequency does not fit with standard or interpolated frequencies, the radio flux value for operating frequency should be obtained by the second interpolation. On www.ips.gov.au, radio flux density is presented also by daily plot diagram and hour plot diagram. These two plot diagrams are presented in Figures 3.15 and 3.16.

From the diagram in Figure 3.16, a very important feature is the constancy of the radio flux density over time for a given frequency. Further, the ground station performance is going to be analyzed through experiment based on the Sun as a source.

Here, a Figure of Merit measurement experiment done with the receiving system at the Vienna's LEO satellite ground station is described, applying the same experiment setup as in Figure 3.9 for the desensibilization experiment.

Results: The experiment was executed several times. Here are used data taken on January 24, 2004. A power meter is used as a spectrum analyzer connected to the downconverter's IF output. The result is presented in Figure 3.17 (Cakaj et al. 2005b).

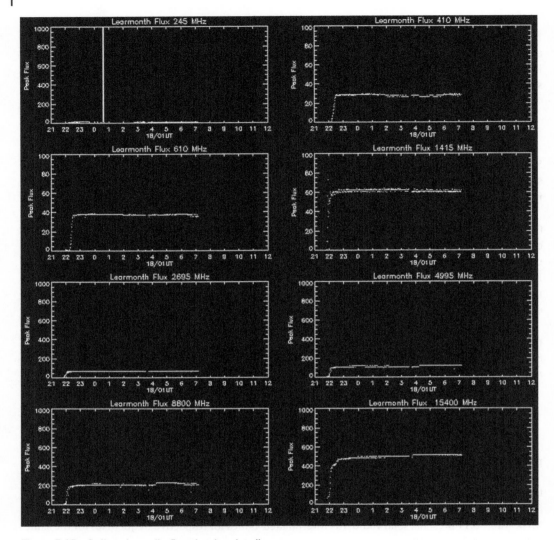

Figure 3.15 Daily solar radio flux density plot diagram.

Table 3.10 Solar radio flux for standard frequencies.

	245 MHz	410 MHz	610 MHz	1415 MHz	2695 MHz	4995 MHz	8800 MHz	15 400 MHz
18.01.07	9	28	37	63	71	119	217	514
24.07.07	9	26	37	56	63	122	225	503
10.08.07	4	24	37	56	69	119	222	511
23.08.07	12	23	36	58	68	123	228	515

Table 3.11 Solar radio flux interpolated values for different frequencies.

	1300 MHz	1540 MHz	2300 MHz	2401 MHz	2790 MHz	5625 MHz	6000 MHz	8000 MHz	10 400 MHz
18.01.07	59.7	64.0	68.9	69.5	73.1	135.0	144.5	196.1	282.3
24.07.07	53.7	56.9	61.2	61.7	65.4	138.7	148.7	203.0	286.1
10.08.07	53.7	57.6	65.5	66.5	71.1	135.6	145.6	199.9	284.7
23.08.07	55.3	59.2	65.4	66.1	70.3	140.0	150.2	205.5	290.8

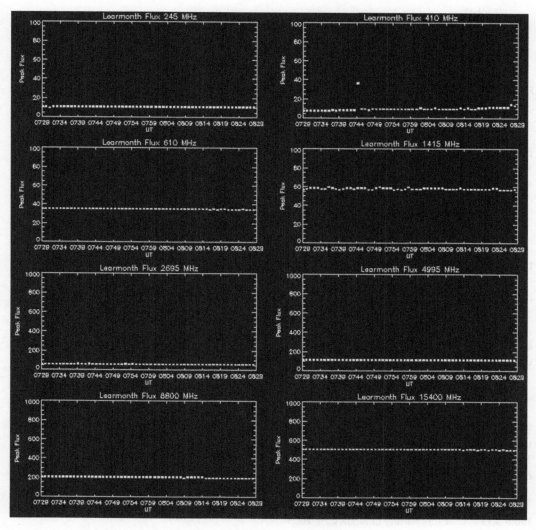

Figure 3.16 Solar radio flux density hour plot diagram.

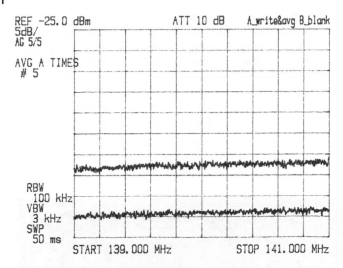

REF -25.0 dBm   ATT 10 dB   A_write&avg B_blank
5dB/
AG 5/5

AVG A TIMES
 # 5

RBW
 100 kHz
VBW
 3 kHz
SWP
 50 ms

START 139.000 MHz   STOP 141.000 MHz

Figure 3.17 Figure of Merit measurement result.

In Figure 3.17 the signal at the bottom represents noise power density when the antenna was pointed to the cold sky. This is proportional to $P_{coldsky}$. Then, we pointed antenna to the Sun. The result of the measurement is presented in Figure 3.17 as an upper signal trace. These two recorded traces are considered for further calculations. The average difference is 11.5 dB.

$$Y(dB) = 10Log\left(\frac{P_{sun}}{P_{coldsky}}\right) = 11.5dB, [Y = 14.125] \tag{3.45}$$

Data for solar radio flux density are provided from the Learmonth Observatory in Australia for the date of January 24, 2004, are presented in Table 3.12 (Cakaj et al. 2005b).

Since the downlink operational frequency is $f = 2232$ MHz and this is not among standard frequencies for Sun flux measurement, the interpolation between fluxes at frequencies at 1415 and 2695 MHz is done. The value for solar radio flux density F_{sun} at frequency 2232 MHz is: $F_{sun} = 91sfu = 91 \cdot 10^{-22} W/m^2 Hz$ (Cakaj et al. 2005b). Substituting Boltzmann's constant $k = 1.38 \cdot 10^{-23} W/HzK$, $f = 2232 \cdot 10^6 Hz$ light's velocity $c = 3 \cdot 10^8 m/s$, Y and F_{sun} at Eq. (3.44), yields Figure of Merit expressed in decibels per degree Kelvin:

$$G/T_S = 14.4dB/K \tag{3.46}$$

The value from link budget calculations ranges between 13.3 dB/K and 15.4 dB/K from Eq. (3.21), where the first value is with atmospheric attenuation ($A = 1$ dB) and the second value represents the ideal case without medium attenuation. The result of measurement done on January 24, 2004, is 14.4 dB/K, and it is within the calculated range. This can be considered as an indication of proper receiving system performance (Cakaj and Malaric, 2007a). The measurements with results

Table 3.12 Solar flux density of January 24, 2004.

f	245 MHz	410 MHz	610 MHz	1415 MHz	2695 MHz	4995 MHz
F_{sun}	12	26	42	74	100	150

presented are executed using the spectrum analyzer R3271 with frequency range from 100 Hz to 26.5 GHz.

Conclusion: The experimental and mathematical results agree. The values extracted from the experiment fall within a range of mathematical calculations for the best and the worst propagation case. This confirms the proper functionality of the LEO satellite ground station.

References

Cakaj, S. (2009). Rain attenuation impact on performance of satellite ground stations for low earth orbiting (LEO) satellites in Europe. *International Journal of Communications, Networks and System Sciences (IJCNS)* 2 (6): 480–485.

Cakaj, S. and Malaric, K. (2006a). Rain attenuation at low Earth orbiting satellite ground station. In: *IEEE, Proc. 48th International Symposium ELMAR 2006 focused on Multimedia Systems and Applications*, 247–250. Zadar, Croatia.

Cakaj, S. and Malaric, K. (2006b). Composite noise temperature at low Earth orbiting satellite ground station. In: *International Conference on Software, Telecommunications and Computer Networks*, 214–217. Split, Croatia: SoftCOM, IEEE.

Cakaj, S. and Malaric, K. (2007a). Rigorous analysis on performance of LEO satellite ground station in urban environment. *International Journal of Satellite Communications and Networking* 25 (6): 619–643.

Cakaj, S. and Malaric, K. (2007b). Isolation measurement between uplink and downlink antennas at Low Earth Orbiting satellite ground station. In: *IEEE Proceedings 19th International Conference on Applied Electromagnetics and Communications*, 24–26. Dubrovnik, Croatia: ICECom.

Cakaj, S. and Malaric, K. (2008). Downlink performance comparison for low Earth orbiting satellite ground station at S-band in Europe. In: *27th IASTED International Conference on Modeling, Identification and Control (MIC)*, 55–59. Innsbruck, Austria.

Cakaj, S., Keim, W., and Malaric, K. (2005a). Intermodulation by uplink signal at low Earth orbiting satellite ground station. In: *IEEE 18th International Conference on Applied Electromagnetics and Communications*, 193–196. Dubrovnik, Croatia: ICECom.

Cakaj, S., Keim, W., and Malaric, K. (2005b). Sun noise measurement at low Earth orbiting satellite ground station. In: *IEEE International Symposium Focused on Multimedia Systems and Applications*, 345–348. Zadar, Croatia: ELMAR.

Cakaj, S., Keim, W., and Malaric, K. (2007). Desensibilization measurement at LEO satellite ground station. In: *IASTED, 26th International Conference on Modelling, Identification and Control*, 36–39. Austria: Insbruck.

Cakaj, S., Kamo, B., Enesi, I., and Shurdi, O. (2011). Antenna Noise Temperature for Low Earth Orbiting Ground Stations at L and S Band. In: *The Third International Conference on Advances in Satellite and Space Communications*, 1–6. Budapest, Hungary: SPACOMM, IARIA.

Elbert, B. (1999). *Introduction to Satellite Communication*. Norwood: Artech House Inc.

Flagg, R. (2003) Determination of G/T" SETI League Inc, Publication Department, AH6NM, New Jersey, USA.

Hoch, J.G. (1979). 'Bestimmung der Empfmdlichkeit von Emphfangsanlagen mitels sonnenrauschen', UKW berichte 4/79. *Numberg* 194–200.

Keim, W., Kudielka, V., and Scholtz, L.A. (2004). A scientific satellite ground station for an urban environment. In: *International Conference on Communications Systems and Networks*, 280–284. Marbella, Spain: IASTED.

Monstein, C. (2002) How to determine antenna temperature in solar radio flux astronomy, HB9SCT, CH-8807, Freienbach, Switzerland. http://www.monstein.de/astronomypublications/ta_form/ta_form.pdf (14.08.2007)

Morgan, M. (2018). Determination of Earth Station Antenna G/T Using the Sun or the Moon as an RF Source. In: *32nd Annual AIAA/USU Conference on Small Satellites*, 1–9. Utah State University, Logan, Utah: https://digitalcommons.usu.edu/cgi/viewcontent.cgi?article=4128&context=smallsat.

Rezaei, P. and Hakkak, M. (2005). Evaluation of interaction effect between LEO ground station antennas. In: *IEEE 18th International Conference on Applied Electromagnetics and Communications*, 197–200. Dubrovnik, Croatia: ICECom.

Saunders, R.S. (1993). *Antennas and Propagation for Wireless Communication System*. Sussex: Wiley.

Sklar, B. (2005). *Digital Communication*, 2e. New Jersey: Prentice Hall PTR.

Zee, E.R. and Stibrany, P. (2002). The MOST microsatellite: a low-cost enabling technology for future space science and technology missions. *Canadian Aeronautics and Space Journal* 48 (1): 43–51.

4

Horizon Plane and Communication Duration

Seen from the satellite ground stations, the horizon planes are categorized in three levels as *ideal*, *practical*, and *designed* one. Each one of these provides the different communication duration between the ground station and the appropriate satellite. To understand the correlation among these horizon planes we first need to examine the principles of the satellite's tracking, which follows as the first section of this chapter.

4.1 LEO Satellite Tracking Principles

Kepler's laws in conjunction with Newton's gravitational law completely explain the motions of planets around the Sun. Satellites movements around the Earth are governed by the same laws (the first Kepler law). From these laws stem the general equation known as the *motion equation* of the satellite's movement within its orbit around the Earth (Maral and Bousquet 2002):

$$\frac{d^2 \vec{r}}{dt^2} + \frac{\mu}{r^3} \vec{r} = 0 \tag{4.1}$$

where $d^2\vec{r}/dt^2$ is acceleration vector of the mass m in the given coordinate system, $\vec{r}$ is vector from the center of the Earth of the mass M to the body with mass m (mass of the satellite), and r is distance between bodies with masses M and m. Finally, μ is the gravitational parameter, and it is: $\mu = GM = 3.986x10^{14} \, \text{m}^3/\text{s}^2$. The solution to this equation describes the trajectory of satellites' movement around the Earth.

The complete solution of Eq. (4.1) is not so simple, and it is not the focus of this book. At this stage, it is enough to say that the solution of this equation is a trajectory within a plane and is shaped as an ellipse, with a maximum extension from the Earth at the *apogee* (r_a) and the minimum at the *perigee* (r_p), as given in Figure 4.1 (Roddy 2006).

The mathematical expression for the appropriate solution is given as:

$$r = \frac{p}{1 + e \cos \theta} \tag{4.2}$$

where r is the distance of any point on the trajectory from the Earth's center (determining the position of the satellite), p is a geometrical constant termed as *conic parameter,* which determines the width of the conic at focus of an ellipse, e is the eccentricity, which determines the type of conic section, and θ is the angle between the r and the point on the conic nearest the focus (*perigee*), determined as *true anomaly.*

Ground Station Design and Analysis for LEO Satellites: Analytical, Experimental and Simulation Approach,
First Edition. Shkelzen Cakaj.
© 2023 The Institute of Electrical and Electronics Engineers, Inc. Published 2023 by John Wiley & Sons, Inc.

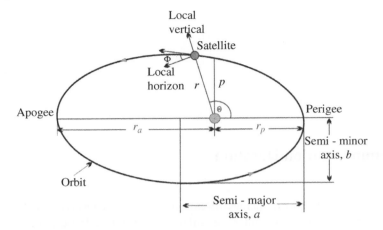

Figure 4.1 Major parameters of an elliptical orbit.

The eccentricity (e) in Eq. (4.2) is the ratio of difference to sum of apogee (r_a) and perigee (r_p) radiuses as in Eq. (4.3).

$$e = \frac{r_a - r_p}{r_a + r_p} \tag{4.3}$$

For the satellites communications that we are concerned with – low Earth orbiting (LEO) satellites – the orbits are circular. In this case is:

$$r_a = r_p \Rightarrow e = 0 \Rightarrow r = p \tag{4.4}$$

Finally, each circular orbit is characterized by its radius from the Earth's center noted by r, and the orbital altitude H is interrelated with its radius r as:

$$r = H + R_E \tag{4.5}$$

where $R_E = 6,378$ km is the Earth's radius. Applying Kepler's laws, we get the satellite's velocity v in the circular orbit:

$$v = \sqrt{\frac{\mu}{r}} \tag{4.6}$$

where r is the orbital radius and $\mu = M \cdot G = 3.986 \cdot 10^5 \text{km}^3/\text{s}^2$ is constant, as a product of Earth's mass and gravitational Earth's constant (Maral and Bousquet 2002; Roddy 2006). The orbital period is expresses as:

$$T = 2\pi\sqrt{\frac{r^3}{\mu}} \tag{4.7}$$

and the number of the daily passes (n) is the ratio of sideral day ($T_{Sideral} = 23$ hours 56 minutes 4.1 seconds) over the orbital period:

$$n = \frac{T_S}{T} \tag{4.8}$$

It must be pointed out that the satellite's velocity in the circular orbit is constant and exclusively depends on the satellite's altitude. So, the LEO altitude is the determining velocity factor for LEO satellites, and consequently for their period and the number of daily passes. For illustration, consider the MOST satellite with data given in Table 1.7, where $H = 830$ km (at Table 1.7 there is a difference between perigee and apogee around of 15 km, but since its impact is so low, the altitude is approximated to 830 km). Applying $H = 830$ km at Eqns. 4.6–4.8 leads to: $v = 7.43$ km/s, $T = 6085$ s (101,4 minutes or 1 hours 41.4 minutes), and $n = 14.21$. This means that this satellite will have 14 full passes and partly a fifteenth one due to one sideral day. Under the ideal conditions, this orbit will keep the velocity and orbital time as unchanged. The main further question is where this orbit is positioned in space related to the Earth's center. This is defined by space orbital parameters (Maral and Bousquet 2002; Roddy 2006; Maini and Agrawal 2011).

Parameters that describe the position of the satellite in space are known as *space orbital parameters,* which is schematically presented in Figure 4.2 (Richharia 1999; Roddy 2006).

Generally, the orbit is elliptic, as given in Figure 4.1, but under specific condition of the zero eccentricity, the orbit becomes circular, which is the case for LEO satellites – the focus of this book. Thus, we'll be talking only about circular orbits. The main question is, where in space is this circle? Geometrically, the circle lies on (belongs to) an *orbital plane*. This orbital plane is somehow positioned in space. To determine the position of the orbital plane in space, two references are applied: equatorial plane and vernal equinox.

Thus, *the position of the orbital plane in space* is specified by means of two parameters – the *inclination i* and the *right ascension of the ascending node* Ω. Inclination i represents the angle of the orbital plane with respect to the Earth's equator. The right ascension of the ascending node Ω defines the location of the ascending and descending orbital crossing points (these two nodes make a *line of nodes*) with respect to a fixed direction in space. The fixed direction is vernal equinox (see Figure 4.2). Vernal equinox is the direction of line joining the center of the Earth and the Sun on the first day of spring (known as spring equinox or vernal equinox) (Richharia 1999; Roddy 2006).

Further, the position of the orbit within an orbital plane (determined by *inclination i* and the *right ascension of the ascending node* Ω) has to be defined. Normally, an infinite number of orbits can be laid within an orbital plane. So, the orientation of the orbit in its plane is defined by the *argument of perigee ω*. This is the angle, taken positively from 0° to 360°, in the direction of the satellite's motion,

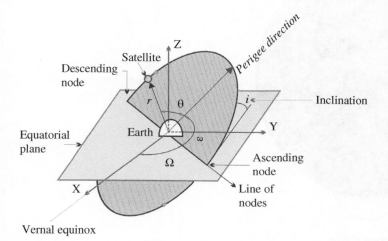

Figure 4.2 Space orbital parameters.

between the direction of the ascending node (line of nodes) and the direction of perigee (Richharia 1999; Roddy 2006).

Finally, the position of the satellite in orbit is determined by the angle θ, called the *true anomaly*, which is the angle measured positively in the direction of a satellite's movement from 0° to 360°, between the direction of perigee and the position of the satellite (see Figure 4.2) (Richharia 1999; Roddy 2006).

From the previous discussion we conclude that the satellite position in its orbit is always known in space and time, where it is orbiting exactly the same path during its whole "life" (normally under ideal conditions – without disturbances). To communicate with this flying body, from the ground station the satellite must be followed, or better saying should be permanently tracked. For these purposes the appropriate software is needed.

Before that, let us discuss the process of establishing the lock/unlock between a ground station and the satellite, which process should be automatically followed by the appropriate software. Let us assume a low Earth circular orbit where the satellite is orbiting. Let us suppose the satellite is in fly mode as undisturbed and it is simply controlled by the Earth's gravity. Also, let us assume a point on Earth, to serve as a ground station (GS) established to communicate with the satellite "Sat" in Figure 4.3. Ground stations can be locked for communications with LEO satellites only when the satellite is in their visibility region.

The satellite and the appropriate ground station cannot communicate until the ground station is covered by the satellite's footprint, given a circle in Figure 4.3 that represents the satellites coverage zone. When the satellite appears at ground station horizon plane, theoretically under zero elevation, the ground station and the satellite should be locked and interchange data, respectively communicate. This is known as acquisition of the satellite (*AOS*). This communication is graphically represented by the line interconnecting the ground station and the satellite as in Figure 4.4. The ground station and the satellite stay connected until the satellite disappears from the ground

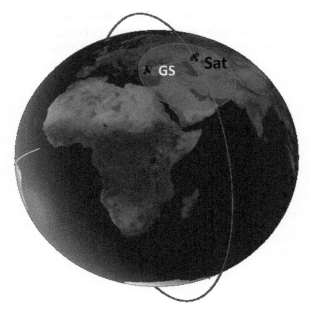

Figure 4.3 Satellite and the ground station.

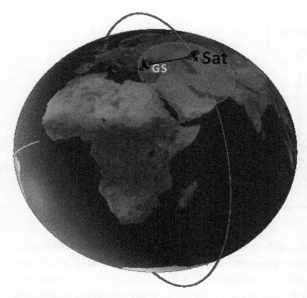

Figure 4.4 Lock is established.

station's horizon, theoretically again under zero elevation, and consequently unlocked or disconnected with the ground station, known as loss of satellite (*LOS*). The time taken, from lock to unlock, determines the communication duration between the ground station and the satellite (Cakaj 2009; Cakaj and Kamo 2018).

This process to be automized by the appropriate software, the software should be fed with the satellite space orbital parameters (known as Kepler elements), which in fact determines the position of the satellite in space at any time. The Keplerian elements determine the exact position of the satellite in space (Keplerian Elements Tutorial-AMSAT 2020). Based on these elements, the satellite's position in space can be predicted using mathematical calculations. Thus, as inputs tracking software takes Keplerian elements (www.celestrak.com) and calculates the actual position of the satellite.

The most common satellite ground stations functions are organized into five main categories, as follows (Landis and Mulldolland 1993):

- *Telemetry processing functions*. These are planned to receive data and all other useful information including video, sent by the satellite, and to process them accordingly to the dedicated mission.
- *Tracking functions*. Examples include orbit determination, altitude determination, current and future position of the satellite with respect to the ground station.
- *Command and control functions*. These are supposed to command and control the events related to satellite. This subsystem controls the satellite, and it is constantly communicating with satellite.
- *Database functions*. These are responsible for storing and retrieving the useful data. For processing the information data is stored for later analysis.
- *Data analysis*. For the data gathered from satellite; this could be done for future studies and scientific diagnostics.

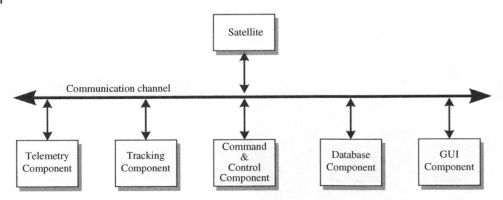

Figure 4.5 Diagram of the satellite ground control station software.

Following these functionalities, the ground station shall comprise a number of respective main modules and components. These modules shall communicate with each other through a common communication channel. Thus, the overall software which enables functionality of the ground station through this communication channel is divided in the following software components, given in Figure 4.5 (Bhatia et al. 2003). Graphical user interface (GUI) component facilitates the operation of the system.

Generally, considering the above components, the software for communication between the ground station and LEO satellite is organized in three main parts:

- The communication software
- The tracking software
- The monitoring ground station software

The communication software provides access to the satellite and controls the communication between the ground station and satellite. It is responsible for setting up a lock/unlock between a satellite and ground station, then to control downloaded data flow. Thus, the command-and-control component, telemetry processing component, and database component are within communication software. This program has connections to all the hardware components.

The tracking software enables the tracking of the satellite. Before the communication to be established it is necessary to "acquire" a satellite. One method of satellite's acquisition is to program the ground station's antenna to perform a scan around the predicted position of the satellite. The automatic tracking system is switched on when the received signal strength is sufficient to lock the tracking receiver. After acquisition a satellite needs to be tracked continuously. This function is performed by the automatic tracking system managed by *tracking software*. As inputs tracking software takes Keplerian elements (www.celestrak.com) and calculates the actual position of the satellite. Information from real-time tracking includes satellite location, satellite, velocity, azimuth, and elevation (pointing or look angles), etc. (Bhatia et al. 2003). Tracking software provides real-time tracking information, usually associated with a map. Maps are presented in these main styles:

- View from space
- Rectangular map
- Radar map

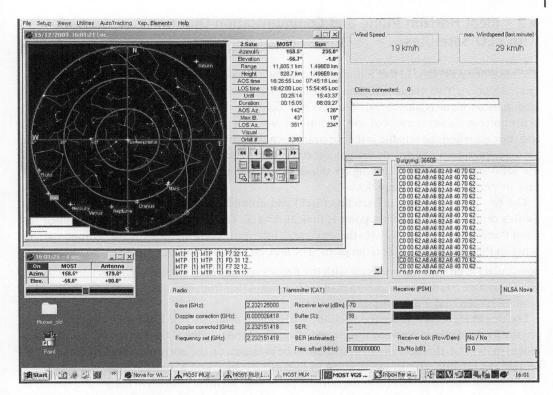

Figure 4.6 Screenshot of tracking software.

The most used style for supervising the satellite is radar map, what is used also throughout this book. For illustration in Figure 4.6 is given a screenshot of the tracking software applied for tracking the MOST satellite from the Vienna ground station (Keim et al. 2004).

The monitoring ground station software is necessary to ensure that the ground station is working properly within normal parameters. This monitoring software includes a number of tests and self-diagnostic modules. If a problem is detected at any ground station device it must either change the parameters for this device for correct operation or shut down the respective device and bring the station into a safe mode. The ground station software should be developed and upgraded in accordance with system's mission.

In order to completely describe the movement of the satellite in space only a few parameters are required to be defined. These are known as space orbital parameters which in fact define:

- The position of the orbital plane in space
- The location of the orbit in orbital plane
- The position of the satellite in the orbit
- The shape of orbit

Tracking software is connected to the antenna's rotator controller unit and moves the antenna when the satellite is above the horizon. Program is arranged on that way; such it moves the antenna few minutes in advance in the waiting position in order to be ready for satellite's acquisition.

4.2 Ideal Horizon Plane and Communication Duration with LEO Satellites

The ideal horizon plane in fact represents the visibility region under 0° of elevation angle from the ground station. Two events, AOS and LOS, under elevation of 0°, geographically determine the *ideal horizon plane* (Cakaj 2009). The first one identifies the case when the satellite appears just at the horizon plane to be locked and to communicate with the ground station (user) and the second one the case when the satellite just disappears from the horizon plane, being unlocked and there is no more communication with the ground station. Further, elaboration explains main features of the ideal horizon plane including its dimension. Looking at another perspective, Figure 1.21 may be presented as in Figure 4.7.

The two points indicate the satellite (SAT) and ground station (*P*), and then the third is the Earth's center. The subsatellite point is indicated by *T*. The horizontal line passing through the point P and having angle of ε_0 with the segment *d* (interconnecting satellite with the ground station) is in fact the ground station horizon plane. The *horizon plane* is considered as a tangent plane (perpendicular) at ground station vector with the Earth's center ($\overrightarrow{R_E}$). Theoretically, each point on the Earth's surface has different horizon plane, what means that from each point on the Earth, the LEO satellite is seen under different look angles.

Looking at Figure 4.7, when the satellite for a moment belongs to the horizon plane when $\varepsilon_0 = 0$, physically the distance between the ground station and the satellite is the longest. As the satellites moves in its orbit, that movement in space may be projected on the horizon plane, as a curve representing the satellite path in space, given in Figure 4.8 (Cakaj and Malaric 2007a).

The "radar map" mode displays the horizon plane with accurate position of the ground station at the center. The perimeter of the circle is the horizon plane, with the North on the top ($A_Z = 0°$), then at the East ($A_Z = 90°$), South ($A_Z = 180°$), and West ($A_Z = 270°$). Three circles represent different elevation 0°, 30°, and 60°. At the center the elevation is $\varepsilon_0 = 90°$, given in Figure 4.8. The curved line represents the satellite path, at each point determined by azimuth and elevation seen from the ground station, what in fact described the satellite path seen from the ground station. Since each ground station has its horizon plane, from each station the satellite is seen differently, having different curve at the horizon plane of the radar map, for the same path of the satellite. Also, from the same ground station, for the satellite under the same orbital parameters for the next satellite path will not be seen under the same look angles as a previous path, since for the time taken for the first satellite path, the Earth rotates under the orbit, consequently changing relative position to the orbit, so the satellite will be seen under different look angles due to the next satellite path.

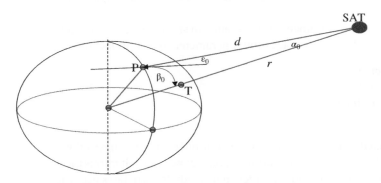

Figure 4.7 Ground station geometry.

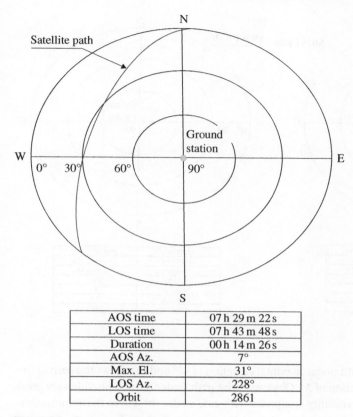

AOS time	07 h 29 m 22 s
LOS time	07 h 43 m 48 s
Duration	00 h 14 m 26 s
AOS Az.	7°
Max. El.	31°
LOS Az.	228°
Orbit	2861

Figure 4.8 Radar map display.

Parameters which determined the satellite path in space, projected as a curve on horizon plane, given as a table under Figure 4.8, are:

AOS_{time} – Acquisition of the satellite (time)
LOS_{time} – Loss of the satellite (time)
AOS_{Ax} – Acquisition of the satellite (Azimuth)
LOS_{Az} – Loss of the satellite (Azimuth)
Max-El – Maximal elevation

Ideal communication duration between the ground station and the appropriate satellite is theoretically defined as the difference in time since the satellite's appearance on the ground station horizon plane (under $\varepsilon_0 = 0°$) and the time when the satellite is lost from the ground station horizon plane (under $\varepsilon_0 = 0°$) expressed as (Cakaj and Malaric 2007a):

$$Duration_{Ideal} = AOS_{time} - LOS_{time} \qquad (4.9)$$

The duration is different for each pass depending exclusively on the *Max-El* as a typical parameter for each different satellite path. The higher the maximal elevation (*Max-El*), the longer is the satellite path above the ground station, so the longer visibility, and consequently the longer communication duration, as illustrated in (Figure 4.9 a,b) as records at Vienna ground station tracking the MOST satellite (Cakaj and Malaric 2007a).

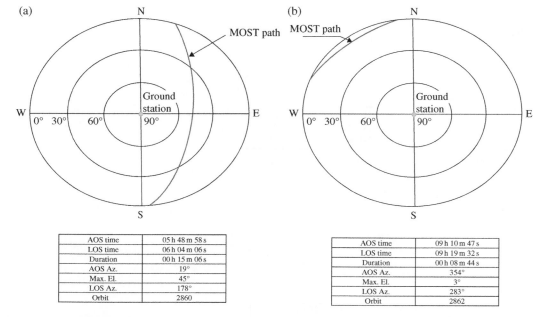

AOS time	05 h 48 m 58 s
LOS time	06 h 04 m 06 s
Duration	00 h 15 m 06 s
AOS Az.	19°
Max. El.	45°
LOS Az.	178°
Orbit	2860

AOS time	09 h 10 m 47 s
LOS time	09 h 19 m 32 s
Duration	00 h 08 m 44 s
AOS Az.	354°
Max. El.	3°
LOS Az.	283°
Orbit	2862

Figure 4.9 Max-El illustration.

Figure 4.9a shows the satellite path under maximal elevation of 45° and (b) shows the path of the same satellite under maximal elevation of 3°. Obviously, the path under (a) will provide very good communication because of longer visibility compared with case under (b) where the communication will be very difficult because of natural barriers at too low elevation.

Ideal horizon plane is defined as a large circle seen from the ground station at the center, under zero elevation $\varepsilon_0 = 0°$ for all azimuths (0°–360°). Its dimension (wideness) will be expressed by its radius or diameter. How much is it?

The distance d from the ground station to the satellite in Figure 4.7, is given by Eq. 1.56, as:

$$d = R_E \left[\sqrt{\left(\frac{H + R_E}{R_E} \right)^2 - \cos^2 \varepsilon_0} - \sin \varepsilon_0 \right] \tag{4.10}$$

or if the elevation expressed as a function of the distance, it is:

$$\sin \varepsilon_0 = \frac{H(H + 2R_E) - d^2}{2dR_E} \tag{4.11}$$

By definition the ideal horizon plane lies at elevation zero ($\varepsilon_0 = 0°$), and considering that the distance d is measured from the ground station positioned at the center of the appropriate horizon plane, then stems out the radius of the circle – ideal horizon plane as:

$$d_{max} = \sqrt{H(H + 2R_E)} \tag{4.12}$$

Finally, the diameter of the ideal horizon plane is d_{max}, consequently the width of the ideal horizon plane is

$$D = 2d_{max}. \tag{4.13}$$

4.3   The Range and Horizon Plane Simulation for Ground Stations of LEO Satellites

Idea: The dimension of the horizon plane depends on satellite's orbital altitude (Eq. 4.12). Here the idea is to conclude what happens with horizon plane by changing the altitude. This leads to conclusion about the communication duration between the ground station and LEO satellites under different altitudes.

Method: The simulation approach is applied. For simulation purposes are considered altitudes from 600 km up to 1200 km, applying the simulation step of 100 km. The elevation is changed in steps of 10°. For these altitudes applying Eq. (4.12) it is calculated the range from a hypothetical ground station and the satellite assumed above the ground station at the altitude H. These data are presented at Table 4.1 and graphically in Figure 4.10 (Cakaj et al. 2011a).

From Figure 4.10 it is obvious that the shortest range occurs at 90° elevation, since the satellites appears perpendicularly above the ground station, and the longest occurs at 0° elevation, also representing the radius of the ideal horizon plane for determined altitude H. The range under the lowest elevation angle represents the worst link budget case, since that range represents the maximal possible distance between the ground station and the satellite.

Result: Considering the slant range under elevation at 0° as the radius of the ideal horizon plane circle, for different values of altitude H, in Figure 4.11, are presented ideal horizon planes for different altitudes. These ideal horizon planes may be considered as circles with diameter from around 6000–8000 km (Cakaj et al. 2011a).[1]

Table 4.1   Slant range between the ground station and the satellite.

Orbital altitude [km]	H [km]	H [km]	H [km]	H [km]	H [km]	H [km]	H [km]
Max-El (ε_0)	600 Range	700 Range	800 Range	900 Range	1000 Range	1100 Range	1200 Range
0°	2830	3065	3289	3504	3708	3900	4088
10°	1942	2180	2372	2577	2770	2955	3136
20°	1386	1581	1765	1947	2120	2287	2453
30°	1070	1234	1392	1549	1701	1849	1996
40°	886	1027	1164	1302	1436	1567	1698
50°	758	883	1005	1128	1248	1366	1486
60°	680	794	905	1018	1129	1238	1348
70°	636	742	847	954	1058	1160	1266
80°	697	707	809	908	1012	1113	1214
90°	600	700	800	900	1000	1100	1200

---

1 On 2016, NASA Marshall Space Flight Center, published a progress report about propulsion system for LEO satellites titled as: *Feasibility study for near term demonstration of laser-sail propulsion from the ground to low Earth orbit,* by authors E. Montgomery, L. Jonson and H. D. Thomas. On page 5, subsection 6. Slant Range, line 10 is written: "*Figure 3 was calculated for any generic ground site using the **method of Cakaj** [16].*"{https://ntrs.nasa.gov/api/citations/20160012066/downloads/20160012066.pdf}

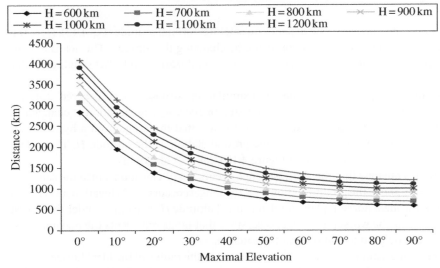

Figure 4.10 Slant range between the ground station and the satellite for different altitudes.

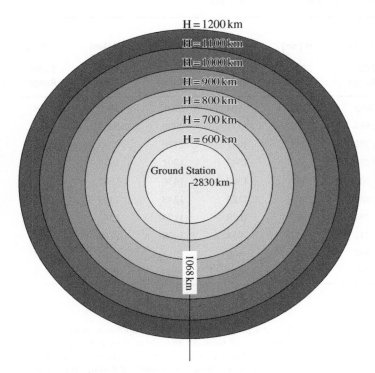

Figure 4.11 Ideal horizon planes.

Conclusion: For ground stations dedicated to communicate with LEO satellites, the ideal horizon plane can be considered as a flat circle with diameter of around 6000–8000 km. Through simulation, it is confirmed that the horizon plan expands as satellite's altitude increases, consequently providing longer communication between satellite and the appropriate ground station.

4.4 Practical Horizon Plane for LEO Ground Stations

Communication via satellite begins when the satellite is positioned in the desired orbital position. Ground stations can communicate with LEO satellites when the satellite is in their visibility region. The duration of the visibility and so the communication duration varies for each satellite pass at the ground station. For low-cost LEO satellite ground stations in urban environment, it is a big challenge to ensure communication down to the horizon. Natural barriers hinder communication at low elevation angles. Thus, motion (appearance) detection of the satellite above natural barriers enables the *practical* horizon to be determined, which usually differs from the ideal horizon (Cakaj 2009).

Idea: The further idea is to identify and clarify the difference in between ideal and practical horizon. How practical horizon affects the communication duration! After emphasizing the appropriate difference, will be defined the concept of the designed horizon plane.

Method: Mathematical, graphical, and on-site records. The satellite pass over the ground station (user) is associated by three characteristic events, happening in this time order. The first one, when the satellite appears just at the horizon plane – defined as the AOS event – establishing the communication with the user station. Theoretically, this happens at elevation of 0° (ideal horizon). The second event is when the satellite achieves the *maximal elevation* (Max-El) of the appropriate satellite pass over the ground station. The third event happens when the satellite disappears from the horizon plane, known as the LOS event, happening theoretically at ideal horizon, elevation 0°. These events in the Figure 4.12 are denoted as AOS, Max-El, and LOS (Cakaj 2021).

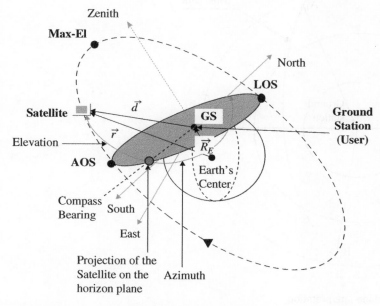

Figure 4.12 The satellite's path characteristic events.

Each satellite pass over the user station is characterized by its *Max-El*, seen from the user's station (Max-El event). For the satellite pass from the event AOS to the event Max-El, the elevation increases (up to Max-El), and from the event Max-El to the event LOS, the elevation symmetrically decreases. Obviously, the higher the angle at the event of Max-El, the satellite path is longer, and the inverse as lower the angle of Max-El, the satellite path is shorter. The shorter satellite path (lower Max-El) enables shorter communication with the ground station (user) and the longer satellite path (higher Max-El) enables longer communication. Finally, as a conclusion stems that the communication duration between the satellite and the user depends on Max-El under which the ground station the appropriate satellite is seen.

Under Section 4.3 it is confirmed that the diameter of the ideal horizon plane is in range of 6,000–8,000 km, which means that in Figure 4.12 the distance from GS to AOS or from GS to LOS is in range of 3,000–4,000 km, for different azimuths. Within these distances, natural barriers can prevent communication to be established under 0° elevation. For better explanation the Figure 4.13 is given.

Theoretically, based on Kepler's laws, the communication between the satellite and the ground station should be established at point A ($Az = 19°$ under 0° elevation) and communication should be lost at point B ($Az = 178°$ under 0° elevation) at Figure 4.13. No contact to the satellite could be established under the 0° elevation, because of natural or artificial barriers. Practically contact between the satellite and ground station is established at point A_1 and lost at point B_1 (both under elevation of few degrees). Point A_1 determines azimuth and elevation when the lock between the ground station and satellite are locked, and B_1 determines azimuth and elevation when the ground station and satellite are unlocked, under real-time practical circumstances. Thus, points A and B belong to the ideal horizon plan; otherwise A_1 and B_1 belong to the practical horizon.

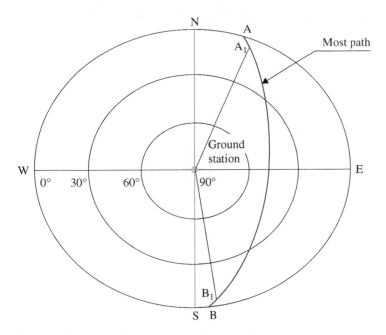

Figure 4.13 Practical horizon plane interpretation.

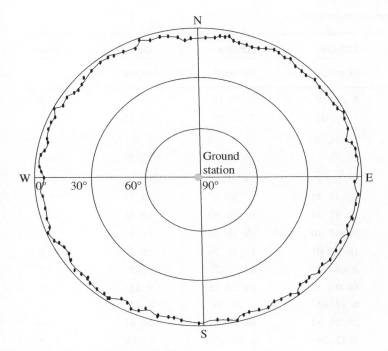

Figure 4.14 Practical horizon plane.

If it is considered that the whole horizon is in the azimuth range of 0°–360°, in any direction of the horizon plane the natural barriers will differ; consequently, so will the acquisition and loss elevation (points A_1 and B_1). Different points for acquisition and loss for different azimuths create the broken circle, as presented in Figure 4.14. The broken circle represents the practical horizon impacted by natural barriers. The practical horizon is not a flat plane. Each edge point of the broken circle is the practical elevation of acquisition or the loss of the satellite. Usually, the lock is established and lost in average at elevation angles of 1°–4° (Cakaj 2009; Cakaj and Malaric 2007a).

The inner line related to the largest circle is in fact is the practical horizon. It is obvious that the practical horizon is always shorter than the ideal one, reflecting on shorter communication time between the satellite and the ground station.

Results: During my time in Vienna for analytical purposes, I recorded around 3000 MOST satellite passes. Table 4.2 shows results of the communication duration for the ideal acquisition and ideal loss of satellite under 0° elevation. This is known as expected communication duration under ideal circumstances. These data are extracted from the tracking software. For the same passes are recorded data under practical circumstances, given in Table 4.3. Just for clarification, lock at practical horizon corresponds to acquisition at ideal horizon, and unlock at practical horizon corresponds to loss at ideal horizon.

For clarification purposes, following are two typical cases, orbits under orbit number 2467 and 2795, both of them are shaded in the tables.

Orbit 2467: This orbit has Max-El of 6° and theoretical expected communication duration is 9 minutes:38 seconds (Table 4.2). But the practical communication lasted only 5 minutes: 0 seconds since the lock and unlock between the ground station and the satellite is established under 4° of elevation (Table 4.3). This means loss on communication time of 4 minutes:38 seconds.

Table 4.2 Expected (Ideal) communication duration.

Date	Orbit	AOStime	LOStime	Duration	Max-El
dd:mm:yy	xxxx	hh:mm:ss	hh:mm:ss	mm:ss	°
05.12.03	2235	5: 31: 39	5: 47: 10	15: 21	84.0
15.12.03	2378	7: 21: 05	7: 33: 23	12: 18	14.0
21.12.03	2467	13: 36: 00	13: 45: 38	9: 38	6.0
13.01.04	2795	16: 11: 18	16: 26: 47	15: 29	86.0
20.01.04	2893	14: 00: 22	14: 11: 27	11: 05	11.0
25.01.04	2965	15: 41: 33	15: 56: 47	15: 14	56.0
28.01.04	3007	14: 45: 14	14: 58: 50	13: 36	22.0
28.01.04	3009	18: 07: 10	18: 18: 41	11: 31	9.0
03.02.04	3093	16: 09: 01	16: 24: 30	15: 29	90.0
08.02.04	3160	8: 08: 24	8: 17: 27	9: 03	8.0
15.02.04	3263	14: 02: 10	14: 13: 25	9: 15	10.0
18.02.04	3302	8: 17: 03	8: 25: 20	8: 17	5.0
20.02.04	3336	18: 24: 34	18: 38: 41	14: 07	23.0
27.02.04	3427	3: 43: 08	3: 55: 43	12: 35	13.0

Table 4.3 Measured data for lock/unlock time.

Date	Orbit	Lock time	Unlock time	AEl	LEl
dd:mm:y	xxxx	hh:mm:s	hh:mm:s	°	°
05.12.03	2235	5: 32: 32	5: 47: 00	3.0	2.0
15.12.03	2378	7: 22: 45	7: 32: 43	4.0	2.5
21.12.03	2467	13: 38: 20	13: 43: 20	4.0	4.0
13.01.04	2795	16: 11: 20	16: 26: 00	2.0	2.0
20.01.04	2893	14: 01: 00	14: 10: 00	2.0	3.5
25.01.04	2965	15: 42: 00	15: 56: 00	2.0	2.0
28.01.04	3007	14: 46: 00	14: 57: 00	3.0	4.5
28.01.04	3009	18: 08: 00	18: 17: 00	3.5	4.0
03.02.04	3093	16: 10: 00	16: 25: 00	2.0	2.5
08.02.04	3160	8: 10: 00	8: 16: 00	3.5	2.5
15.02.04	3263	14: 03: 00	14: 12: 00	3.0	3.5
18.02.04	3302	8: 20 : 00	8: 23: 00	4.0	3.0
20.02.04	3336	18: 25: 00	18: 37: 00	3.0	2.5
27.02.04	3427	3: 44: 00	3: 54: 00	3.0	3.0

Orbit 2795: This orbit has Max-El of 86° and theoretical expected communication duration is 15 minutes:29 seconds (Table 4.2) But the practical communication lasted 14 minutes: 40 seconds since the lock and unlock between the ground station and the satellite is established under 2° of elevation. This means loss on communication time is 49 seconds.

Conclusion: For recorded satellite passes the lock is established and lost in average at elevation angles of 1°–4°. Obviously, practical horizon differs from the ideal one, at least for 1°–4° degrees of elevation in average, because of natural barriers. This is confirmed based on records at Vienna satellite ground station. Thus, the ideal communication duration exclusively depends on orbital Max-El. But under real circumstances because of natural barriers, the real communication duration depends on lock/unlock elevation between the satellite and the appropriate ground station. As higher orbital Max-El and as lower lock/unlock elevation, the loss in communication duration is less – consequently, better communication performance. This aspect is further quantified.

4.5 Real Communication Duration and Designed Horizon Plane Determination

Idea: From the previous section, we can use measurements to confirm the difference between software expected (ideal) communication duration and real-time (practical) communication duration between the ground station and the satellite. The idea is this difference to be quantified and applied due to the designing of the LEO satellite ground station. For these purposes, we implement the *time efficiency factor* (T_{eff}) (Cakaj and Malaric 2007a).

Method: Mathematical and experimental approach is applied. To quantify these variations in communication duration (comparing expected-ideal to real communication time), we define a parameter called T_{eff} as:

$$T_{eff} = \frac{LOCK_{time} - UNLOCK_{time}}{AOS_{time} - LOS_{time}} \qquad (4.14)$$

where AOS_{time} and LOS_{time} are, respectively, expected (ideal) acquisition and loss time of the satellite. The difference of these two parameters is defined by Eq. (4.9) as an *ideal communication duration* between the satellite and the ground station. But, practically, because of natural barriers, the lock with satellite is established with time delay. This is called lock time, designated as $LOCK_{time}$. Also, by the same reasons, the satellite communication loses a little bit in advance. This is called unlock time, designated as $UNLOCK_{time}$. Thus, the practical (real) duration of communication is, in fact, defined as a difference between lock and unlock time, expressed by Eq. (4.15) as:

$$Duration_{real} = LOCK_{time} - UNLOCK_{time} \qquad (4.15)$$

T_{eff} represents the ratio of the real communication duration to the theoretical-expected communication duration. This factor will be used for interpretation of communication duration due to low elevation angles. Applying (4.14) and Eq. (4.15) yields that real communication duration is:

$$LOCK_{time} - UNLOCK_{time} = T_{eff}(AOS_{time} - LOS_{time}) \qquad (4.16)$$

$$Duration_{real} = T_{eff}(AOS_{time} - LOS_{time}) \qquad (4.17)$$

Finally, at smaller time efficiency factors the real communication is more decreased compared with ideal communication duration. The appropriate difference is expressed as:

$$\Delta Duration = (AOS_{time} - LOS_{time}) - (LOCK_{time} - UNLOCK_{time}) \qquad (4.18)$$

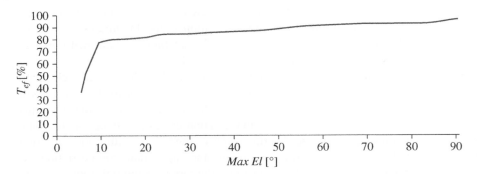

Figure 4.15 Time efficiency factor dependency on maximum elevation angle.

The communication efficiency due to the time variations as well as the usefulness of low elevation passes is analyzed for the satellite ground station within space observation project "MOST." For tracking the satellite, a tracking mechanism and software are used.

Results: The records of 3000 MOST satellite passes seen over Vienna ground station are analyzed in the format as data provided in Tables 4.2 and 4.3. Normally, these data are also interpolated, avoiding illogical cases. From these analyses is extracted the following diagram of T_{eff} on dependence of Max-El (Cakaj and Malaric 2007a).

Figure 4.15 shows T_{eff} in percentage as a function of Max-El for the data from 3000 satellite passes. The diagram in Figure 4.15 has a breaking point at Max-El of around 10°. For Max-El higher than 10° the time efficiency factor is keeping a trend of linearity starting at 80% toward 100%, but for Max-El lower than 10°, T_{eff} rapidly falls, causing a huge difference in the window of possible communication. From the communication point of view, we are much safer in communication above breaking point. This fact should also be considered during the link budget. Above the breaking point, we avoid all uncertainties due to natural barriers; thus, designed horizon plane is located at this point or above, greatly increasing the likelihood of unbroken communication. The designed horizon plane is determined by its designed minimal elevation as given in Figure 4.16 (thicker circle under the broken circle).

Conclusion: Each LEO satellite ground station is characterized by its designed horizon plane. The designed horizon plane is defined by the minimal elevation considered for the acquisition and loss of satellite. For different purposes of the satellite systems, the minimal elevation value for the designed horizon plane ranges from 5° to 30° (Essex 2001). For example, the NOAA's search and rescue ground stations uses 5° for the designed horizon plane (Cakaj et al. 2010). Breaking point of time efficiency factor is directly correlated with elevation of the designed horizon plane. Finally, the higher time efficiency factor, the shorter the difference on ideal and real communication duration is, which means better usage of the system.

4.6 Ideal and Designed Horizon Plane Relation in Space

Two events, AOS and LOS, under elevation of 0°, designated as AOS (0°) and LOS (0°) geographically determine the *ideal horizon plane*. The virtual line connecting points in space when AOS and LOS happen at elevation of 0° determine the *ideal horizon plane wideness* (*IHPW*) presented in Figure 4.17. The width of the horizon plane depends on satellite's orbital altitude (Cakaj et al. 2011a).

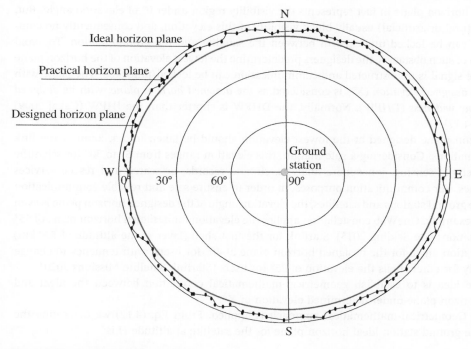

Figure 4.16 Designed horizon plane.

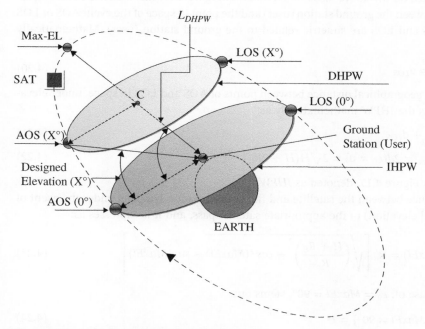

Figure 4.17 Ideal and designed horizon plane.

The ideal horizon plane in fact represents the visibility region under 0° of elevation angle. But, barriers (natural or artificial) usually block visibility at this elevation, and consequently no communication can be locked/unlocked in between the satellite and the ground station. To avoid the problem of such obstacles, the designers predetermine the lowest elevation of the horizon plane at which the signal is unobstructed and communication can be locked/unlocked. The plane with appropriate *designed elevation* (X°) is considered as the *designed horizon plane*, with its *designed horizon plane wideness* (DHPW). Normally, the DHPW is shorter than the IHPW (Cakaj 2009) (see Figure 4.17).

The horizon plane designed at the lowest elevation should be taken into account at the link budget estimations. Considering a safe margin, this elevation ranges from 5° to 30° for scientific purposes, remote sensing, ionosphere analysis, climate records, or search and rescue services LEO satellites. For communication purposes, in order to ensure safe and reliable communication between the ground stations and satellites, the elevation angle of the designed horizon plane is even higher. For example, OneWeb constellation applies the elevation for designed horizon plane of 55° for users' stations (De Selding 2015). Starlink for the first shell (layer at the altitude of 550 km) applies elevation angle for the designed horizon plane at 40° for users, with tendency to change it or to apply for other shells the elevation of 25° for users (Starlink Satellite Missions 2020).

Idea: The idea is to establish geometrical-mathematical correlation between the ideal and designed horizon plane under determined elevation of $X°$.

Method: Geometrical-mathematical approach is applied. From Eq. (4.12) we determine the radius of the ground station ideal horizon plane for the satellite at altitude H is:

$$d_{max} = d(\varepsilon_0 = 0) = \sqrt{H(H + 2R_E)} \qquad (4.19)$$

This range under the lowest possible elevation angle $\varepsilon_0 = 0$, is the maximal possible distance between the ground station (user) and the satellite of the altitude H, and consequently the worst link budget case. Based on the above discussion related to Figure 4.17, this distance in fact represents the distance between the ground station (user) and the point in space of the event AOS or LOS at $\varepsilon_0 = 0$, since AOS and LOS are simetric related to the ground station (user). Mathematically expressed, it is:

$$d_{max} = d_{AOS} = d_{LOS} \qquad (4.20)$$

Finally, the virtual geographical distance between points of AOS and LOS in space, under elevation $\varepsilon_0 = 0$, represents the IHPW mathematically as:

$$IHPW = d_{AOS} + d_{LOS} \qquad (4.21)$$

$$IHPW = 2d_{max} = 2d(\varepsilon_0 = 0) = 2\sqrt{H(H + 2R_E)} \qquad (4.22)$$

and schematically as Figure 4.17, denoted as *IHPW*.

The minimal distance between the satellite and the ground station happens under the event of the Max-El (maximal elevation) of the appropriate satellite pass, and it is expresses as:

$$d_{min}(\varepsilon_0 = MaxEl) = R_E \left[\sqrt{\left(\frac{H + R_E}{R_E}\right)^2 - \cos^2(MaxEl)} - \sin(MaxEl) \right] \qquad (4.23)$$

and for the special case of: $\varepsilon_0 = MaxEl = 90°$, stems out:

$$d_{min}(\varepsilon_0 = MaxEl = 90°) = H \qquad (4.24)$$

Now, let us assume that the designed horizon plane is defined by the lowest elevation $\varepsilon_{0D} = X°$ (ε_{0D} is added index D, referred to the "designed") (see Figure 4.17). For the Starlink satellites constellation, for the first shell (layer at the altitude of 550 km), it is applied elevation angle for the designed horizon plane of 40° for users, but since it could be considered changeable, for further analytic elaboration it is considered $X° = [25, 30, 35, 40]$.

The maximal distance between the satellite and the ground station (user) for the ($\varepsilon_{0D} = X°$) is determined by $\varepsilon_{0D} = X°$, as:

$$d_{max}(X°) = d(\varepsilon_{0D} = X°) \tag{4.25}$$

Finally, the satellite's movement within its own orbit related to the designed horizon plane at $\varepsilon_{0D} = X°$ is as follows, interpreted through Figure 4.17.

The satellite appears at the ideal horizon due to the event AOS (0°) but not locked with the ground station (user) until the event AOS ($X°$) at elevation $\varepsilon_{0D} = X°$ when it is locked with the ground station (user) and having the slant range $d_{max}(X°)$, (for this case, the maximal range is achieved at designed $X°$). Thus, even when the satellite is above the user there is no lock (no communication) between the satellite and the user from the event AOS (0°) to the event AOS ($X°$). The lock is established just at AOS ($X°$). Further the satellite flies higher toward the event of the maximal elevation at $\varepsilon_0 = MaxEl$ attaining the slant range d_{min}, and then orbiting down toward the event LOS ($X°$) at elevation $\varepsilon_{0D} = X°$ having again the slant range $d_{max}(X°)$, unlocked from the ground station (user), and still flying above the ground station (user) but with no communication until the satellite disappears at event LOS (0°). This is the communication cycle between the satellite and the ground station (user). It should be emphasized that the minimal slant range (under 90°) as d_{min} remains the same, independent of the designed elevation, since the user does not change its position by the designed elevation, only the designed horizon plane moves up remaining always parallel with the ideal horizon plane. Thus, the line connecting AOS ($X°$) and LOS ($X°$) represents DHPW, given in Figure 4.17 and mathematically expressed as:

$$DHPW = d_{AOS(X)} + d_{LOS(X)} \tag{4.26}$$

where $d_{AOS(X)}$ and $d_{LOS(X)}$ are measured related to the point C in Figure 4.17. Point C is the up projection of the user station at the designed horizon plane.

The next step is to find out what the designed horizon plane wideness is compared to the ideal one and how far apart they are from each other. Looking from the ground station (user), the *DHPW* is the base of virtual up napped cone with the apex exactly at the ground station (user). The DHPW is in fact the diameter of the base of the appropriate virtual cone. Solving the triangle ground station (user) – AOS ($X°$) – center of *DHPW* (C) in Figure 4.17 (Cakaj 2021), yields:

$$DHPW = 2\,d_{AOS(X)} = 2d(\varepsilon_{0D} = X°)\sin(90 - X) = 2d(\varepsilon_{0D} = X°)\cos X \tag{4.27}$$

Denoting by L_{DHPW} the parallel distance of the designed horizon plane from the ideal one (see Figure 4.17), from the same triangle yields out that the L_{DHPW} is:

$$L_{DHPW} = d(\varepsilon_{0D} = X°)\cos(90 - X) = d(\varepsilon_{0D} = X°)\sin X \tag{4.28}$$

Finally, (4.27) and (4.28) express the geometrical features of the designed horizon plane, its dimension and its distance from the ideal horizon plane.

Results: For three shells, Starlink planned constellation, with shell's altitudes of $H = 550$ km, 1110 km, 340 km and designed elevation $\varepsilon_{0D} = 25°, 30°, 35°, 40°$ of the designed horizon plane at users' sites (ground station), based on Eqns. (4.19), (4.22), (4.27), (4.28), we can calculate the

IPHW and DPHW for the appropriate elevation. Further the vertical parallel distance (L_{DHPW}) of the designed DHPW related to the ideal horizon plane (IHPW) are calculated and given in Table 4.4 (Cakaj 2021).

Conclusion: This section is closed with graphical presentation of the shaded case (actually under operation) from Table 4.4, to create a feeling about dimensions (values) and relation in space, given in Figure 4.18. These values are completely in accordance with results given under section 4.3, where the ideal horizon plane is simulated. The area of the designed horizon plane determines the zone where the ground station and satellite may communicate with each other.

Table 4.4 IHPW, DHPW and L_{DHPW} for the different shells.

| The designed horizon plane elevation (ε_{0D}) | The first shell H = 550 km | | The second shell H = 1110 km | | The third shell H = 340 km | |
| | IHPW (km) = 5405.8 | | IHPW (km) = 7833.9 | | IHPW (km) = 4206.8 | |
	DHPW (km)	L_{DHPW} (km)	DHPW (km)	L_{DHPW} (km)	DHPW (km)	L_{DHPW} (km)
25°	2045.2	476.3	3744.1	871.8	1323.6	308.3
30°	1718.4	496.0	3226.7	931.5	1094.3	315.9
35°	1465.5	512.6	2795.1	977.7	926.8	324.2
40°	1240.1	519.7	2405.1	1007.8	775.9	325.3

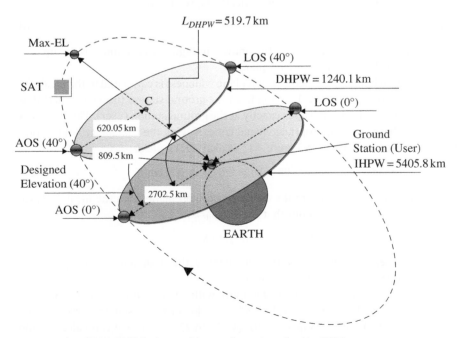

Figure 4.18 IHDW, DHPW, d_{max} and L_{DHPW} dimensions for H = 550 km.

4.7 Savings on Transmit Power through Designed Horizon Plane at LEO Satellite Ground Stations

Each ground station is characterized by its own ideal horizon plane. To avoid natural barriers, that plane must be modified to the designed one, defined by minimal elevation in order to avoid natural obstacles. Designed horizon plane implementation implies that less power will be transmitted from the satellite. The major loss in communication between the LEO satellite and the ground station is the free space loss. Free space loss varies since the distance from the ground station to the satellite varies over time, usually compensated through variable satellite transmit power toward the downlink.

Idea: Small and lightweight satellites (micro and nano) have flexible payload architectures with variable satellite EIRP and reconfigurable coverage in order to maximize satellite mission performance (Acquaroli and Morelli 2005). On-board processing models with reconfiguration and self-configuration are considered, also (Perlaza et al. 2006). The further idea is the mathematical elaboration and confirmation that through the implementation of the designed horizon plane instead of an ideal one, the transmit EIRP (equivalent isotropic radiated power) from the satellite to the ground station could be saved, and permanently keeping unchanged receiving performance at the ground station (compared to the case of ideal horizon plane). This implies mandatory agile power transmitter at the satellite, transmitting the needed power according to the designed horizon plane of the ground station.

Method: Mathematical and simulation approach applied. The goal is to prove the EIRP savings at the satellite, by keeping the receiving signal margin flat (unchanged, constant) at the station, through the implementation of the designed horizon plane, instead of an ideal one. In order to obtain the constant downlink margin at the receiver (constant signal to noise ratio), all over the time, for analytical and simulation purposes, we consider the typical LEO altitudes from 600 km to 1200 km. For these altitudes, under different elevations, are simulated and calculated the EIRP savings. For simulation purposes the increment steps of 200 km for altitudes and of 5° for elevation are applied. The power savings simulation is presented, proving the power savings by designed horizon implementation compared to the ideal one (Cakaj and Kamo 2018).

The downlink margin (DM) is defined as:

$$DM = (S/N)_r - (S/N)_{rqd} \tag{4.29}$$

where r indicates expected signal to noise ratio to be received at receiver, and rqd means required signal to noise ratio requested by customer, as in advance defined performance criteria. The positive value of DM is indicating that the link is closed (communication is established), since what is received is better than required.

The further simulation elaborates the EIRP (equivalent isotropic radiated power) savings at the satellite by implementing the designed horizon plane instead of ideal one at the ground station, due to the keeping constant downlink receiving margin. Thus, the idea behind this approach is the assumption of keeping the constant downlink margin over different distances between the LEO satellite and the appropriate ground station. For simulation purposes, we consider altitudes of 600 km, 800 km, 1000 km, and 1200 km as typical LEO altitudes.

$$DM = (S/N)_r - (S/N)_{rqd} = constant \tag{4.30}$$

Considering that the required $(S/N)_{rqd}$ level is determined in advance by user requirements, the result must be:

$$(S/N)_r = constant \tag{4.31}$$

and since it is

$$(S/N)_r = (S/N_0)_r - B \tag{4.32}$$

where B is determined receiving system bandwidth, and the simulation is applied for the ground system with already determined bandwidth, it should be:

$$(S/N_0)_r = constant \tag{4.33}$$

The LEO satellite path over the ground station is characterized by typical events above the ground station, the first one, when the satellite is seen due to the longest distance from the ground station (for ideal horizon plane under 0° elevation: AOS (0°) and LOS (0°), or for designed horizon plane under $X°$ elevation: AOS($X°$) and LOS ($X°$)), and the second, when the satellite is seen perpendicularly from the ground station (under 90° elevation). To fulfill the condition expressed by Eq. (4.33), the lowest EIRP is required under 90° elevation and the highest EIRP is required for ideal horizon plane under 0°, and in between for designed horizon plane under $X°$. The lowest required EIRP, for both cases, ideal or designed horizon plane, is the same since in both cases the range is the same as H (satellite's altitude).

Under random elevation $x°$ and neglecting other loss, the range equation becomes as:

$$\left(\frac{S}{N_0}\right)_{(\varepsilon_0 = x)} = EIRP_{(\varepsilon_0 = x)} - L_{S(\varepsilon_0 = x)} + \left(\frac{G}{T_S}\right) + 228.6 \tag{4.34}$$

(G/T_S) it is not indexed by elevation since the environmental factors are approximated for all cases and technical equipment parameters do not depend on elevation. Then, (4.34) becomes:

$$\left(\frac{S}{N_0}\right)_{(\varepsilon_0 = 0)} = \left(\frac{S}{N_0}\right)_{(\varepsilon_0 = x)} = \dots = \left(\frac{S}{N_0}\right)_{(\varepsilon_0 = X)} = \left(\frac{S}{N_0}\right)_{(\varepsilon_0 = 90)} \tag{4.35}$$

Being aware that the index 0 refers to ideal horizon plane (I), index x refers to any random point of practical horizon (P), X refers to designed horizon plane (D) and index 90 refers to the perpendicular distance (H) above the ground station, the Eq. (4.34) can be rewritten as:

$$\left(\frac{S}{N_0}\right)_I = \left(\frac{S}{N_0}\right)_P = \left(\frac{S}{N_0}\right)_D = \left(\frac{S}{N_0}\right)_H \tag{4.36}$$

Applying (4.34) will have following four equations:

$$\left(\frac{S}{N_0}\right)_I = EIRP_I - L_{S(I)} + \left(\frac{G}{T_S}\right) + 228.6$$

$$\left(\frac{S}{N_0}\right)_P = EIRP_I - L_{S(P)} + \left(\frac{G}{T_S}\right) + 228.6$$

$$\left(\frac{S}{N_0}\right)_D = EIRP_D - L_{S(D)} + \left(\frac{G}{T_S}\right) + 228.6$$

$$\left(\frac{S}{N_0}\right)_H = EIRP_H - L_{S(H)} + \left(\frac{G}{T_S}\right) + 228.6 \tag{4.37}$$

For the communication between the satellite and the ground station, to keep constant margin, the EIRP transmitted from the satellite toward the ground station should vary, respectively, as follows, for ideal horizon plane, for designed horizon plane, for practical horizon and under the perpendicularity, as follows:

$$EIRP_H < EIRP_D < EIRP_P < EIRP_I \tag{4.38}$$

Applying (4.36) yields out as:

$$EIRP_I - L_{S(I)} = EIRP_P - L_{S(P)} = EIRP_D - L_{S(D)} = EIRP_H - L_{S(H)} \tag{4.39}$$

Eq. (4.39) tells us that for larger free space loss (larger distance), more power must radiate from the satellite in order to keep constant performance. Further concern is the comparison on the power to be transmitted under the circumstances of ideal and designed horizon plane. The cases are indexed under I and D, so will have:

$$EIRP_I - L_{S(I)} = EIRP_D - L_{S\,(D)} \tag{4.40}$$

$$\Delta EIRP = EIRP_I - EIRP_D \tag{4.41}$$

$$\Delta EIRP = L_{S(I)} - L_{S(D)} \tag{4.42}$$

Finally, applying expression for the free space loss, for the case of the ideal horizon (0° elevation) and for the designed horizon plane ($X°$ elevation), then the saving on EIRP transmitted from the satellite, for the designed horizon plane considered under the minimal elevation of $X°$, compared to the power needed for the communication at circumstances under ideal ground station horizon plane (0° elevations) is expressed as (Cakaj and Kamo 2018):

$$\Delta EIRP_X\,(dB) = 20\,log\,\frac{d(\varepsilon_0 = 0)}{d(\varepsilon_0 = X)} \tag{4.43}$$

Results: Applying Eq. (1.56) related to the slant range between the ground station and the satellite, for $\varepsilon_0 = 0°$ and then for $\varepsilon_0 = X = 5°$, 10°, 15°, 20°, 25°, 30°, then substituting the respective values at Eq. (4.43) (see Table 4.5), represent the saving on EIRP for the designed horizon planes determined by different minimal elevation compared to ideal horizon plane. These results are also given in Figure 4.19.

From Table 4.5, it is obvious that for altitude of 800 km, under elevation of 10°, the savings on EIRP are almost 3 dB (2.85 dB), which means that in the designated plane, the power transmitted from the satellite toward the ground station may be approximately a half of the power needed at the ideal horizon plane, keeping the same margin. Figure 4.19 shows that for a determined satellite altitude (H), the power savings due to implementation of designed horizon plane, increases as the minimal elevation increases. On the right angle in Figure 4.19, X denotes the minimal elevation for the designed horizon plane. For the altitude of 800 km, "playing" with elevation on the range of 5–30°, the length of bricks varies from 1.44 to 7.47 dB, in fact representing the range of the power savings by designed horizon plane implementation. Similar savings result for other altitudes.

Table 4.5 Power savings (ΔEIRP).

Orbital altitude [km]		H 600 [km]	H 800 [km]	H 1000 [km]	H 1200 [km]
Horizon plane	Elevation (ε_0)	ΔEIRP [dB]	ΔEIRP [dB]	ΔEIRP [dB]	ΔEIRP [dB]
Designed	5°	1.67	1.44	1.28	1.16
Designed	10°	3.27	2.85	2.53	2.30
Designed	15°	4.61	4.16	3.73	3.39
Designed	20°	6.20	5.40	4.85	4.43
Designed	25°	7.31	6.46	5.83	5.35
Designed	30°	8.44	7.47	6.77	6.22

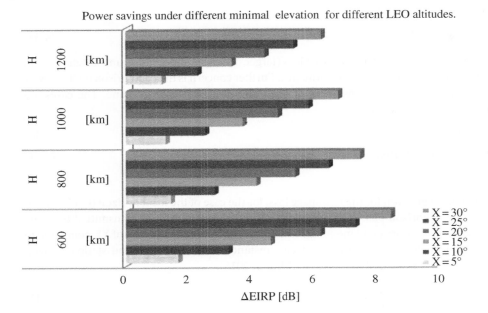

Figure 4.19 Power savings (ΔEIRP).

Conclusion: Both mathematically and through simulation, we can confirm that implementing a *designed horizon plane* has two important benefits. Not only does the higher elevation avoid obstacles caused by natural or manmade barriers but also less power is needed to transmit the signal from the satellite, keeping constant (unchanged) margin. The methodology on designed horizon plane implementation and advantages compared to an ideal horizon plane are given. Calculations about how power level should be managed form the satellite to the ground station under different designed horizon planes are also given.

4.8 Elevation Impact on Signal-to-Noise Density Ratio for LEO Satellite Ground Stations

From previous sections, it is confirmed that the satellite's slant path range related to the ground station varies over time, depending on elevation angle (Eq. 1.56). The shortest range is achieved under maximal elevation of satellite's path above the ground station. Thus, the signal toward the ground station is faced with different free space loss. The variation on free space loss (FSL), under different elevation impacts the receiving ground station performance, respectively the link budget. For the downlink performance, of greatest interest is the receiving system signal-to-noise ratio (S/N) or (S/N_0). (S/N) depends on the last end receiving device bandwidth. To avoid the effect of the last end receiving device bandwidth, we further consider the impact of elevation on signal-to-spectral-noise density ratio (S/N_0).

   Idea: To calculate, simulate, and graphically present the impact of elevation on the signal-to-noise density ratio for LEO satellites seen from the ground station. For analysis, S-band is considered.

Method: Math and simulation approach is applied. To find out the dependence on signal-to-noise density ratio on elevation angle, it is assumed $EIRP = 30 \text{dBW}$ and $G/T_s = 15 dB/K$. For simulation purposes satellite altitudes are considered from 600 km up to 1200 km. Another quantitative factor for comparison purposes is defined, as the maximal difference of signal to spectral noise density ratio under elevation of 90° and 0°, respectively, mathematically expressed, as:

$$\Delta \frac{S}{N_0} (dB/Hz) = \left(\frac{S}{N_0}\right)_{90} - \left(\frac{S}{N_0}\right)_0 \tag{4.44}$$

For the downlink budget calculations, the receiving S/N_0 is the most interesting parameter, expressed by range equation as:

$$\frac{S}{N_0} (dB/Hz) = EIRP - L_S - L_0 + \frac{G}{T_S} + 228.6 \tag{4.45}$$

where $EIRP$ is effectively isotropic radiated power from the transmitter, L_s is free space loss, L_0 is other losses (including atmospheric, polarization, and pointing loss) (G/T_s) is satellite ground station Figure of Merit, and the value 228.6 yields from the Boltzmann's constant. The most influent factor on S/N density ratio is free space loss, which directly depends on elevation angle.

Free space loss (L_S) is the greatest loss in transmitted power due to the long distance between the satellite and a ground station. This loss is displayed as:

$$L_S = \frac{(4\pi d)^2}{\lambda^2} \tag{4.46}$$

where d is the distance (slant range) between the satellite and a ground station, and λ is the signal's wavelength. The free space loss L_S is often convenient to be expressed as function of distance d and signal frequency f, and then L_S is

$$L_S(\varepsilon_0) = \left(\frac{4\pi f}{c}\right)^2 \cdot d^2(\varepsilon_0) \tag{4.47}$$

where $d(\varepsilon_0)$ is represented by Eq. (4.48). Free space loss increases by both, frequency and the distance. ε_0 denotes the elevation angle:

$$d(\varepsilon_0) = R_E \left[\sqrt{\left(\frac{H + R_E}{R_E}\right)^2 - cos^2 \varepsilon_0} - sin \varepsilon_0 \right] \tag{4.48}$$

Usually, LEO satellites operate at S-band (Keim and Scholtz 2006; Cakaj and Malaric 2007a), thus for $f = 2$ GHz, Eq. (4.46) becomes:

$$L_S(\varepsilon_0)[dB] = 98.45 + 20 \log d(\varepsilon_0) \tag{4.49}$$

where $d(\varepsilon_0)$ is represented by (4.48) and is expressed in [km]. The free space loss under considered altitudes of 600, 800, 1000, and 1200 km is presented in Table 4.6 and Figure 4.20 (Cakaj et al. 2014).

See Table 4.6 for LEO satellites at orbit altitudes in range from 600 km up to 1200 km, because of elevation variation, there is a variation on free space loss in average of 12 dB, impacting on link budget calculations and consequently the S/N spectral density ratio. Applying Eq. (4.49) at Eq. (4.45), will get:

$$\frac{S}{N_0} (dB/Hz) = EIRP - L_0 - 20 \log d(\varepsilon_0) + G/T_S + 130.15 \tag{4.50}$$

Table 4.6 Free space loss (L_S) for LEO satellites.

Orbital Altitude [km]	H 600 [km]	H 800 [km]	H 1000 [km]	H 1200 [km]
El (e_0)	L_S [dB]	L_S [dB]	L_S [dB]	L_S [dB]
0°	167.5	168.8	169.8	170.7
10°	164.2	165.9	167.3	168.3
20°	161.3	163.4	165.0	166.2
30°	159	161.3	163.0	164.4
40°	157.4	159.8	161.6	163.0
50°	156.2	158.5	160.4	161.9
60°	155.3	157.6	159.5	161.0
70°	154.8	157.0	158.9	160.5
80°	154.3	156.6	158.6	160.1
90°	154.0	156.5	158.5	160.0

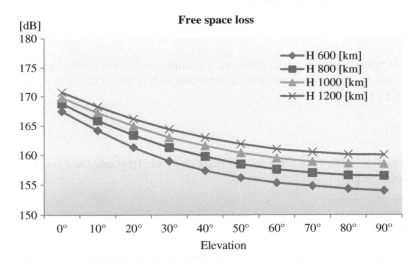

Figure 4.20 Free space loss diagram.

Other losses denoted as L_0 include atmospheric loss, polarization loss, and pointing loss. It is confirmed that atmospheric loss at S-band is not larger than 1 dB in Central Europe (Cakaj and Malaric, 2006a, Cakaj and Malaric 2007a). Since the idea is to consider only impact of elevation on signal-to-spectral noise density ratio for LEO satellite ground stations, and then all other losses are considered as negligible.

It is assumed a hypothetical ground station with composite temperature (including lines and equipment) of $T_{comp} = 70$ K (Cakaj and Malaric 2006b) and receiving antenna with gain of $G = 35$dBi. At S-band LEO satellite ground stations will have very similar performance in central Europe (Cakaj and Malaric 2008) Considering antenna noise temperature in average of 30 K for Central Europe (Cakaj et al. 2011b), the Figure of Merit is:

$$G/T_S(dB/K) = 15 \qquad (4.51)$$

Finally, signal-to-noise density ratio is:

$$\frac{S}{N_0}(dB/Hz) = EIRP - 20 \log d(\varepsilon_0) + 145.15 \qquad (4.52)$$

This equation tells us that the variation on free space loss has to be compensated by EIRP in order to keep constant S/N_0 when LEO satellite is flying at different elevation. Assuming the LEO satellite transmitting EIRP as 30dBW will have:

$$\frac{S}{N_0}(dB/Hz) = 175.15 - 20 \log d(\varepsilon_0) \qquad (4.53)$$

Applying data from Table 4.6 to Eq. (4.53) we get the S/N_0 for different LEO orbit altitudes presented in Figure 4.21.

Further, the maximal difference of S/N_0 under elevation of 90° and 0° is:

$$\Delta \frac{S}{N_0}(dB/Hz) = \left(\frac{S}{N_0}\right)_{90} - \left(\frac{S}{N_0}\right)_0 \qquad (4.54)$$

Results: Applying Eq. (4.53) to Eq. (4.54), for respective values related to elevation 90° and 0° we get maximal difference of signal-to-noise spectral density ratio presented in Table 4.7 (Cakaj et al. 2014).

Table 4.7 confirms that the maximal difference of S/N_0 decreases as the LEO orbital altitude increases. The maximal difference of S/N_0 for different orbital attitudes is around 12 dB on average.

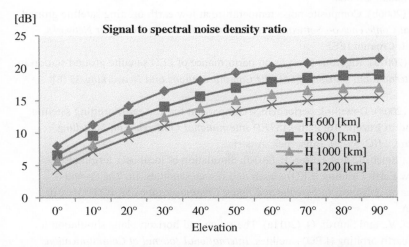

Figure 4.21 Signal-to-spectral noise density ratio for LEO ground stations.

Table 4.7 Maximal difference of signal-to-spectral noise density ratio.

Orbital Altitude [km]	H 600 [km]	H 800 [km]	H 1000 [km]	H 1200 [km]
$\Delta \dfrac{S}{N_0}(dB/Hz)$	13.5	12.4	11.3	11.2

Conclusion: The range variation causes the free space loss changes over the same satellite path depending on the look elevation angle from the ground station. This variation on received signal-to-noise spectral density at ground station caused because of variation on elevation for LEO satellites has to be compensated by EIRP from satellite. For the downlink performance, the receiving signal-to-noise spectral density ratio (S/N_0) is of high importance.

References

Acquaroli, L. and Morelli, B. (2005) Agile payload development and testing of the simplified qualification model. *The European Conference on Spacecraft Structures, Materials and Mechanical Testing,* Noordwijk, The Netherlands 2005, Bibliographic Code: 2005ESASP.581E.56A.

Bhatia, M., Jonson, R., Zalewski, J. (2003) Evaluating performance of real-time software components: satellite ground control station, Case study, Florida Space Institute, Orlando, USA.

Cakaj, S. (2009). Practical horizon plane and communication duration for low earth orbiting (LEO) satellite ground stations. *WSEAS Journal: Transactions on Communications* 8 (4): 373–383.

Cakaj, S. (2021). The parameters comparison of the "Stralink" LEO satellites constellation for different orbital shells. *Frontiers in Communications and Networks-Aerial and Space Networks* 2: 643095.

Cakaj, S. and Kamo, B. (2018). Savings on transmit power through designed horizon plane for LEO satellite ground stations. *Journal of Communications Software and Systems (JCOMSS)* 14 (3): 264–271.

Cakaj, S. and Malaric, K. (2006a). Rain attenuation at low earth orbiting satellite ground station. In: *IEEE, Proceedings of the 48th International Symposium ELMAR 2006 focused on Multimedia Systems and Applications,* 247–250. Zadar, Croatia.

Cakaj, S. and Malaric, K. (2006b). Composite noise temperature at low earth orbiting satellite ground station. In: *International Conference on Software, Telecommunications and Computer Networks, SoftCOM,* 214–217. Split, Croatia: IEEE.

Cakaj, S. and Malaric, K. (2007a). Rigorous analysis on performance of LEO satellite ground station in urban environment. *International Journal of Satellite Communications and Networking* 25 (6): 619–643.

Cakaj, S. and Malaric, K. (2008). Downlink performance comparison for low earth orbiting satellite ground station at S-band in Europe. In: *27th IASTED International Conference on Modelling, Identification and Control, MIC,* 55–59. Innsbruck, Austria.

Cakaj, S., Fitzmaurice, M., Reich, J., and Foster, E. (2010). Simulation of local user terminal implementation for Low Earth Orbiting (LEO) search and rescue satellites. In: *The Second International Conference on Advances in Satellite and Space Communications SPACOMM 2010,* 140–145. Athens: IARIA.

Cakaj, S., Kamo, B., Kolici, V., and Shurdi, O. (2011a). The range and horizon plane simulation for ground stations of low earth orbiting (LEO) satellites. *International Journal of Communications, Networks and System Sciences (IJCNS)* 4 (9): 585–589.

Cakaj, S., Kamo, B., Enesi, I., and Shurdi, O. (2011b). Antenna noise temperature for low earth orbiting satellite ground stations at L and S band. In: *The Third International Conference on Advances in Satellite and Space Communications SPACOMM 2011,* IARIA, 1–6. Budapest, Hungary.

Cakaj, S., Kamo, B., Lala, A., and Rakipi, A. (2014). Elevation impact on signal to spectral noise density ratio for low earth orbiting satellite ground station at S-band. In: *IEEE Science and Information Conference,* 641–645. London.

De Selding, B.P. (2015). Virgin, Qualcomm invest in OneWeb Satellite internet venture. *SpaceNews* ((January 15)).

Essex, A. E. (2001) Monitoring the ionosphere/plasmasphere with low Earth orbit satellites: The Australian Microsatellite FedSat. Cooperative Research Center for Satellite Systems, Department of Physics, La Trobe University, Bundoora. Corpus ID: 127212553.

Keim, W. and Scholtz, L.A. (2006). Performance and reliability evaluation of the S-band, at vienna satellite ground station, talk, IASTED. In: *International Conference on Communication System and Networks*, 5. Palma de Mallorca, Spain.

Keim, W., Kudielka, V., and Scholtz, L.A. (2004). A scientific satellite ground station for an urban environment. In: *International Conference on Communication Systems and Networks, IASTED*, 280–284. Marbella, Spain.

Keplerian Elements Tutorial – AMSAT (2020), https://www.amsat.org/keplerian-elements-tutorial.

Landis, J.S. and Mulldolland, E.J. (1993, 1993). Low-cost satellite ground control facility design. *IEEE, Aerospace & Electronics systems* 2 (6): 35–49.

Maini, K.A. and Agrawal, V. (2011). *Satellite Technology*, 2e. West Sussex: Wiley.

Maral, G. and Bousquet, M. (2002). *Satellite Communication Systems*. Chichester, Sussex: Wiley.

Perlaza, M.S., Hoyos, A.E., and Vera, V.P. (2006). Reconfigurable satellite payload model based on software radio technologies. In: *3rd IEEE International Congress of the Andean Region*, 1–6. Ecuador, ANDESCON.

Richharia, M. (1999). *Satellite communications systems*. McGraw -Hill New York.

Roddy, D. (2006). *Satellite communications*. New York: McGraw Hill.

Starlink Satellite Missions (2020). https://directory.eoportal.org/web/eoportal/satellite-missions/s/starlink (Accessed January 17, 2021).

5

LEO Coverage

5.1 LEO Coverage Concept

The low Earth orbiting (LEO) satellite's coverage area represents the fraction of the Earth's surface from where users have visibility with satellite, can access and establish communication with satellite. Users (stations) on the ground can communicate with LEO satellites when the user (station) is under the coverage area (satellite footprint) as presented in Figure 5.1. Since the satellite is always on the move, the coverage area moves also, so that at various points the user is out of the footprint, and consequently loses communication. Visibility duration and consequently communication duration varies for each LEO satellite pass over the ground station, since LEO satellites move quickly over the Earth, several times during the sideral daytime. The satellite's coverage area is usually expressed as a percentage of the Earth's surface, which is really a low percentage for LEO satellites – just a few percent of Earth's area.

The coverage area of a single satellite is a circular area (Figure 5.1) on the Earth's surface from where the satellite can be seen under an elevation angle equal to or greater than the minimum elevation angle determined by the system/mission requirements. The largest coverage area of a LEO single satellite is achieved under the elevation of $\varepsilon_0 = 0°$, which is not always provided because of natural/artificial barriers under too-low elevation. The single LEO satellite coverage aspects serve as an overture to the global coverage.

Often there is confusion between the coverage area and the ideal/designed horizon plane. The ideal horizon plane is the virtual flat surface laid perpendicularly on Earth's radius vector (Figure 1.21 or 4.12), where the radius vector connects the Earth's center with the ground station within the satellite's coverage area. On the other hand, the coverage area is spherical area on the ground, where each point within that area has its own horizon plane, providing different view of the satellite. Thus, within the satellite's coverage area, each user has its own ideal/designed horizon plane, so each user on the ground has different communication line with the single satellite. The designed horizon plane is simply the parallel with the ideal one distanced from it by the distance (L_{DHPW}) determined by the designed elevation angle (see Eq. 4.28).

The satellite is looking down at its coverage area for the user to be locked and communicate, and on the other side the user is looking up at its own horizon plane for the satellite to be locked for the communication.

Ground Station Design and Analysis for LEO Satellites: Analytical, Experimental and Simulation Approach, First Edition. Shkelzen Cakaj.
© 2023 The Institute of Electrical and Electronics Engineers, Inc. Published 2023 by John Wiley & Sons, Inc.

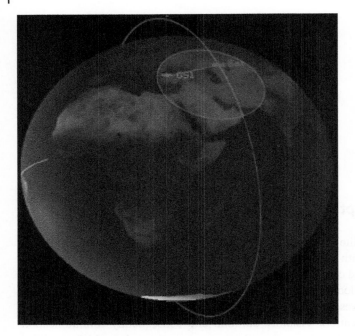

Figure 5.1 The ground station (GS) under the LEO coverage area.

5.2 LEO Coverage Geometry

For the further mathematical calculation, the geometry description is applied. The coverage geometry for LEO satellites is given in Figure 5.2. The LEO satellite is orbiting above the Earth at altitude H. In Figure 5.2 two coverage cases are given: the first one, the largest (full) coverage is under the elevation of $\varepsilon_0 = 0°$ and the second one is under the predetermined elevation, or better expressed

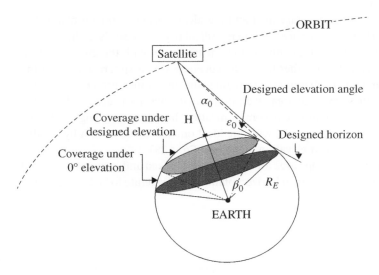

Figure 5.2 LEO coverage area geometry.

under the designed elevation. There are two triangles in Figure 5.2. The largest triangle is related to the largest (full) coverage, which is given by the larger circle. The smaller triangle is related to the LEO coverage area on the Earth's surface under a few degrees of elevation (predefined elevation-designed elevation) given in the figure by the smaller circle. For both triangles: ε_0 is elevation angle, α_0 is nadir angle, β_0 is central angle, and d is the slant range. $R_E = 6371$ km is the Earth's radius. Generally, for triangles, it is:

$$(\varepsilon_0 + 90) + \alpha_0 + \beta_0 = 180 \tag{5.1}$$

Since the ideal horizon plane is always perpendicular with Earth's radius vector, this yields:

$$\varepsilon_0 + \alpha_0 + \beta_0 = 90 \tag{5.2}$$

Furthermore, applying the sinus theorem:

$$\frac{sin\,\alpha_0}{R_E} = \frac{sin\,(90 + \varepsilon_0)}{R_E + H} \tag{5.3}$$

$$sin\,\alpha_0 = \frac{R_E}{R_E + H}\,cos\,\varepsilon_0 \tag{5.4}$$

The full coverage is achieved for $\varepsilon_0 = 0$, and this condition determines the largest nadir angle of the satellite's propagation toward the Earth:

$$\alpha_{0,max} = sin^{-1}\left(\frac{R_E}{R_E + H}\right) \tag{5.5}$$

By definition, the coverage C [%] is the fraction of the Earth's surface covered by the satellite, expressed as the ratio of the satellite coverage area ($SAT_{COVERAGE}$) to the Earth's surface (S_{EARTH}) as:

$$C\,[\%] = \frac{SAT_{COVERAGE}}{S_{EARTH}} \tag{5.6}$$

From (Richharia, 1999; Roddy 2006), or even from elementary geometry it is: $SAT_{COVERAGE} = 2\pi R_E^2\,(1 - cos\,\beta_0)$ and $S_{EARTH} = 4\pi R_E^2$, thus the coverage area by LEO satellite expressed in percentage is:

$$C\,[\%] = \frac{1}{2}\,(1 - COS\beta_0) \tag{5.7}$$

The satellite's coverage area of Earth depends on orbital parameters – more accurately, the coverage as a percentage value or as a surface expressed in square kilometers depends on altitude and elevation angle, but as a position on Earth's surface also depends on inclination.

5.3 The Coverage of LEO Satellites at Low Elevation

The largest coverage area is achieved under elevation of 0°, but the lock and unlock between the satellite and the appropriate ground station under this elevation is too difficult to establish because of the natural/artificial barriers. That is, the line of site between the satellite and the ground station is hindered by natural/artificial obstacles interfering with the communications. To avoid such obstacles at too-low elevation, usually a designed elevation angle is determined. But, under the designed elevation, the coverage area decreases, which leads to the compromise to be applied for the link budget calculations.

Idea: The idea is to draw conclusions about the reliability in communication between the satellite and the ground station under too low elevation. More exactly, it is to define the elevation angle, which provides the safe and reliable communication and its impact on the coverage area width, and consequently, on the range between the satellite and the ground station. The determined range is the crucial component for the appropriate link budget calculations.

Method: Simulation and math calculations are applied. For the coverage area analysis under too-low elevation, the basic geometry between a satellite and ground station is applied, where (α_0) is nadir angle, (β_0) is central angle, and (ε_0) is elevation angle. For simulation purposes, the altitudes from 600 km up to 1200 km are considered. Since the satellite's coverage strongly depends on elevation angle, also for the simulation purposes, it is considered the elevation up to 10°, elevated by steps of 2° (Cakaj et al., 2014). For a given satellite altitude H and a given elevation angle ε_0 should first be calculated α_0, from Eq. (5.4) and then β_0 from Eq. (5.2). Finally, the coverage area expressed in percentage is calculated based on Eq. ((5.7). For altitudes of $H = 600$ km, 800 km, 1000 km, and 1200 km, which are typical low-orbit altitudes, it is simulated and calculated the coverage area for the elevation of (0–10)° by steps of 2°.

Results: Under the previous simulation assumptions, the results are presented in Table 5.1 and Figure 5.3 (Cakaj et al., 2014).

Table 5.1 and Figure 5.3 confirm the decrease of coverage area as elevation angle increases for the already defined altitude H, and the increase of the coverage area as altitude H increases, keeping the elevation fixed.

Finally, Figure 5.4, applying satellite orbit analysis software, presents the case of simulated coverage area for synchronized orbits at an altitude of 600 km for different inclination (few orbits) at elevation of 10°, as the smallest coverage area stems from the simulation previously considered. Small circles in Figure 5.4 represent LEO coverage area on Earth's surface.

Conclusions: Satellite coverage strongly depends on elevation angle. The largest coverage area is achieved under elevation of 0°, but in order to avoid obstacles caused by natural barriers at too-low elevation, usually for the link budget calculations, the minimal elevation angle is determined, which ranges from 2° to 10°.

Through simulation for typical LEO attitudes on range of 600–1200 km at low elevation of 0° to 10°, it is confirmed that the fraction of Earth covered by satellites at appropriate attitudes is 1.69% to 7.95%.

Table 5.1 Coverage areas as a fraction of Earth area.

Orbital Attitude [km]	H 600 [km]	H 800 [km]	H 1000 [km]	H 1200 [km]
Elevation (ε_0)	*Coverage [%]*	*Coverage [%]*	*Coverage [%]*	*Coverage [%]*
0°	4.30	5.60	6.80	7.95
2°	3.63	4.84	5.95	7.08
4°	3.05	4.16	5.21	6.22
6°	2.53	3.49	4.54	5.48
8°	2.08	3.01	3.91	4.75
10°	1.69	2.54	3.38	4.20

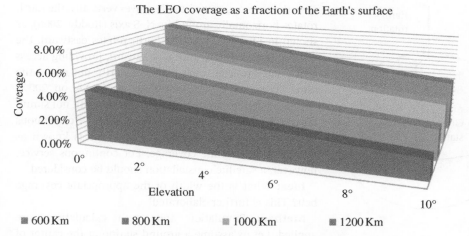

Figure 5.3 Coverage area variation for different altitudes at low elevation.

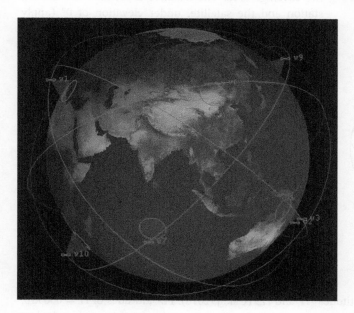

Figure 5.4 LEO coverage area.

5.4 Coverage Belt

The LEO satellite orbit is in principle fixed and the satellite flies strongly keeping its path determined by Kepler elements (if no disturbances). Thus, as the satellite orbits around the Earth, consequently the coverage area on Earth under the satellite, also virtually moves vertically, as presented in Figure 5.5. The Earth's area swept by LEO satellite's coverage during one orbit path is known as a coverage belt as given in Figure 5.6. The width of the appropriate coverage belt depends on satellite altitude. Satellites under the same altitude but under different inclination make different belts, enabling global coverage, but not instant one.

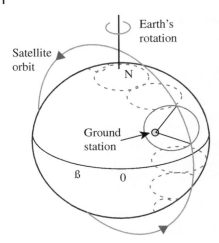

Figure 5.5 Virtual coverage movement.

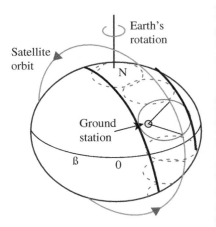

Figure 5.6 Coverage belt.

As the coverage area virtually moves vertically, the Earth rotates horizontally around its N–S axis (Roddy, 2006), as shown in Figure 5.7. As the Earth rotates eastward, the coverage belt virtually moves westward, providing access to the satellite from each point on Earth, but not at the same (provided by author's time at NOAA on 2009). A single satellite can provide global access, but not simultaneously, thus a single one cannot provide continuous real-time services. Single-satellite coverage is known as an *individual* satellite coverage. For continuous service, however, a satellite constellation should be considered.

Idea: What is the width of the appropriate coverage belt? This is further elaborated!

Method: Simulation and math calculations are applied. Let us assume a ground station at the center of the coverage area. The distance d between the ground station and the satellite, under elevation of $0°$ (apply Eq. (1.56)), represents the radius of the largest coverage area's circle. The width of the coverage belt is twice that of the largest radius, expressed as follows:

$$d_{(\varepsilon_0 = 0)} = d_{max} = R_E \left[\sqrt{\left(\frac{H + R_E}{R_E} \right)^2 - 1} \right] \quad (5.8)$$

$$D_{BELT} = 2 d_{max} \quad (5.9)$$

Based on Eq. (5.8) and Eq. (5.9), it is obvious that the satellite coverage belt strongly depends on altitude and elevation angle. The largest one is under elevation of $0°$. To conclude about the coverage belt variation for low orbiting satellites, the simulation for altitudes from 600 km up to 1200 km is further applied, as illustrated by Figure 5.8. Schematically in Figure 5.8 is presented the belt wideness variation for two different satellites SAT1 and SAT2, under different altitudes *H1* and *H2* (Cakaj, 2016). For altitudes of $H = 600$ km, 800 km, 1000 km, and 1200 km, which are typical low-orbit altitudes, we simulate and calculate the coverage belt width for elevations of $(0–10)°$ by steps of $2°$.

Results: The results are presented in Table 5.2 and Figure 5.9, which confirm the decrease of coverage belt width as the elevation angle increases for the already defined altitude H, and the increase of the coverage belt width as altitude H increases under the fixed elevation.

Conclusions: The satellite's coverage belt on the Earth depends strongly on orbit altitude and elevation angle. The widest coverage belt is achieved under elevation of $0°$, but in order to avoid obstacles caused by natural barriers at too-low elevation, usually for the link budget calculations, the minimal elevation angle ranges from $2°$ to $10°$. For the higher elevation consequently the narrower coverage belt is generated. Through simulation for typical LEO attitudes of 600–1200 km, it is confirmed that the coverage belt width covered by satellites at respective altitudes ranges from 5633 to 8177 km.

Figure 5.7 LEO coverage belt.

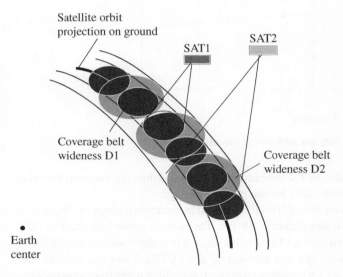

Figure 5.8 Coverage belt of different satellite altitudes.

5.5 LEO Global Coverage

Recently, LEO satellites are oriented on providing real-time continuous services, what implies the necessity for the global coverage and handover process. The global satellite coverage can be considered as an interoperable complementary networking process of multiple satellites organized in a constellation, each of them contributing with its individual coverage (Zong and Kohani, 2019;

Table 5.2 Coverage belt width.

Orbital Attitude [km]	H 600 [km]	H 800 [km]	H 1000 [km]	H 1200 [km]
Elevation (ε_0)	*D* [km]	*D* [km]	*D* [km]	*D* [km]
0°	5633.0	6579.0	7416.0	8177.8
2°	5215.2	6157.2	6991.4	7751.2
4°	4824.4	5760.2	6590.6	7347.2
6°	4463.0	5386.8	6210.0	6959.8
8°	4141.6	5048.8	5859.2	6601.2
10°	3857.4	4745.0	5541.8	6273.6

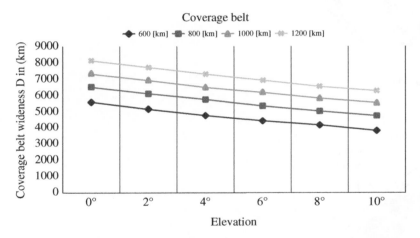

Figure 5.9 Coverage belt variation for different altitudes at low elevation.

Seyedi and Safavi, 2012). Let us first discuss the individual coverage and then the necessity for other satellites involvement (creating constellation) for the global coverage.

The simulated coverage area based on the LEO satellite orbital parameters is given in Figure 5.10. (Cakaj, 2021). The altitude applied for simulation is 800 km. The user on ground (defined as LUT-KOS for simulation purposes) is locked with a LEO satellite since it is under coverage area (satellite footprint). The lock is symbolized with the line connecting the LUTKOS and the satellite SAT (Cakaj, 2010). For simulation purposes and coverage interpretation, four more basic communication points (BC) are given, as BC1, BC2, BC3, and BC4.

Under the case presented in Figure 5.10, only two of them (BC1 and BC4) can be locked with the satellite since they are within satellite's footprint, and two others (BC2 and BC3) are out of the footprint, so with no communications possibility. This means that BC1 and BC4 can communicate with each other via satellite, but not with BC2 and BC3 since the last ones are not accessible by satellite. This is known as *communication within coverage area* (footprint), or as *individual satellite coverage*. As the satellite moves down in its orbit, the coverage area also vertically moves, leaving the ground station (LUTKOS) and two base stations (BC1, BC4) out of the footprint and consequently losing the communication (Cakaj et al., 2014).

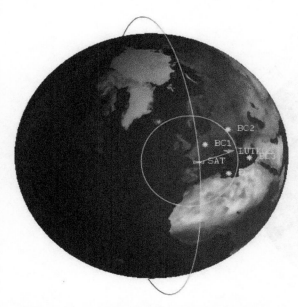

Figure 5.10 Simulated LEO satellite's coverage area.

Let us further suppose the second LEO satellite with the same altitude is orbiting the Earth as shown in Figure 5.10. Since, both have the same altitude, the dimensions of the satellite area are the same, just covering different zones. Let us suppose that coverage areas are adjacent, so the second satellite is covering basic communication points BC2 and BC3. This means that BC2 and BC3 can communicate with each other intermediated by the second satellite, but no one of them with BC1 and BC4 since the last ones are not accessible by the second satellite. If the first and second satellite can communicate with each other, these satellites will enable the communication among all four basic communication points. This is interoperability complementary process by two satellites providing communications within two coverage areas. This concept – step by step, adding more satellites, to build an organized *satellite constellation* – provides global coverage. The Iridium satellite constellation system is among the earliest systems providing global coverage planned to provide worldwide communication services, shown in Figure 5.11 (Jong et al. 2014).

LEO satellites organized in constellations act as a convenient network solution for global coverage and real-time services (Zong and Kohani, 2019; Seyedi and Safavi, 2012). The LEO constellation is a system of LEO identical satellites, launched in several orbital planes with the orbits having the same altitude (single-layer constellation). The satellites move in a synchronized manner in trajectories relative to Earth. Satellites in low orbits arranged in a constellation work together by relaying information to each other and to the users on the ground. Each individual satellite of constellation contributes with its coverage, toward global coverage created by interoperability. The application of LEO satellites organized in a *constellation* is an alternative to wireless telephone networks.

Global coverage should be so smooth that the user does not detect that the signal is being handed over from one satellite to another (Seyedi and Safavi, 2012). Handover and management policies become even more critical under very low elevation because of natural barriers. Handover policies and management are well analyzed (Alyildiz et al. 1999; Papapetrou et al. 2003).

The handover process is highly risky at these low elevations; thus, for complete coverage of the Earth's surface, some overlap between adjacent satellites is necessary so that users on the

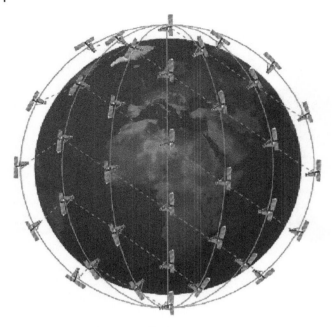

Figure 5.11 Iridium constellation.

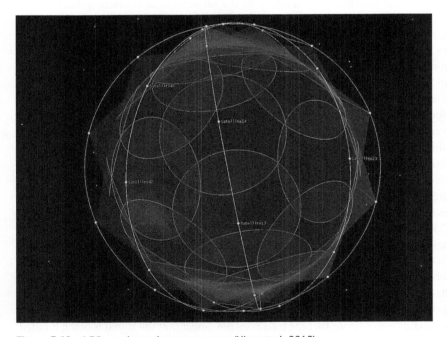

Figure 5.12 LEO overlapped coverage area (Xiao et al. 2018).

ground do not sense any lack of continuity in real-time services. Usually, the coverage areas overlap by few degrees. Overlapping of the satellite constellation is given in Figure. 5.12 (Xiao et al. 2018). The handover process is executed within the overlapped area. The satellites within a constellation are equipped with advanced on-board processing; they can intercommunicate with each other by line of sight using intersatellite links (ISL), applied for the handover process and intersatellite traffic.

When designing a satellite network some decisions such as the selection of the orbit parameters, coverage model, the network connectivity and routing model must be made. Different deterministic models for coverage time evaluation of LEO satellite are developed. Models involve statistical coverage time assessments. The analyses are particularly useful for probabilistic investigation of inter-satellite handovers in LEO satellite networks (Papapetrou et al. 2003). The probability of service interruption and handover mechanism becomes important for the overall system performance (Seyedi and Safavi, 2012).

5.6 Constellation's Coverage – Starlink Case

Technological efforts toward an integrated satellite-terrestrial network stem by the end of the last century, especially with the applications of LEO microsatellites and nanosatellites. Integrated satellite-terrestrial networking toward providing global internet broadband services recently reflects the highest research scientific and industry interests worldwide. Active satellite projects related to an integrated satellite-terrestrial communication network include Iridium constellation with 66 satellites (Cochetti, 2015), OneWeb constellation with 648 satellites (De Selding, 2015; Pultarova and Henry, 2017), and Telesat with 117 spacecrafts in its constellation (Foust, 2018). The Federal Communications Commission (FCC) has approved Amazon's plan to launch 3236 spacecraft in its Kuiper constellation, but the largest undertaking is the Starlink satellite constellation, coming from SpaceX, a private US company majority owned by Elon Musk (Starlink, 2020; Starlink Satellite Missions, 2020).

The Starlink constellation is planned to be organized in three spatial shells, each of them with several hundred small-dimensioned and lightweight LEO satellites specially designed to cover the entire Earth with broadband services through a satellite-terrestrial integrated network, connecting ground stations with 4,425 LEO satellites.

As of October 24, 2020, 893 satellites were accommodated in orbits of altitude of 550 km under different inclinations, determining the first Starlink orbital shell. Two next generations are planned for altitudes of 1110 and 340 km, completing the appropriate infrastructure of three Starlink satellite shells, toward ubiquitous presence of broadband internet services (Cakaj, 2021).

Nearly 12 000 satellites are planned to be deployed, organized in three orbital shells, as follows (Starlink, 2020; Starlink Satellite Missions, 2020):

- The first shell: 1440 satellites in a 550 km altitude.
- The second shell: 2825 satellites in a 1110 km altitude.
- The third shell: 7500 satellites in 340 km altitude.

The first shell of 1440 satellites will be into 72 orbital planes of 20 satellites each, and others later by intention to be completed by 2024 and to provide real-time broadband services (Starlink, 2020; Starlink Satellite Missions, 2020). Figure 5.13 presents a train of SpaceX Starlink satellites visible and captured by satellite tracker Marco Langbroek in Leiden, the Netherlands, on May 24, 2019 (Rao, 2019).

Idea: These three satellite's orbital shells manifest different coverage on Earth because of their different altitudes. Here we want to compare three shells under different elevation.

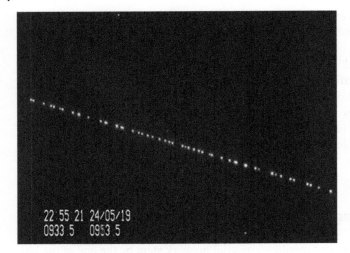

Figure 5.13   Starlink satellite train.

Method: The coverage area calculation under section 5.2 is applied. For the Starlink constellation, considering three shells at altitudes $H = 550$, 1110, and 340 km for the full coverage under elevation of $\varepsilon_0 = 0$ and for the designed elevation $\varepsilon_{0D} = 25°$, $30°$, $35°$, $40°$ based on Eq. (5.2), (5.4) are calculated nadir angle $\alpha_0°$ and central angle $\beta_0°$ as the first step for the coverage calculation for all three Starlink shells. Then, applying the above calculated $\beta_0°$ and considering three shells at altitudes $H = 550$, 1110, and 340 km for the full-coverage elevation of $\varepsilon_0 = 0$ and for the designed elevation $\varepsilon_{0D} = 25°$, $30°$, $35°$, $40°$ based on Eq. ((5.7), the respective coverage for all three Starlink shells is calculated (Cakaj, 2021).

Results: The mathematical outcomes are presented in Table 5.3 and Table 5.4.

Conclusions: The too-low fraction of the Earth's surface covered by LEO satellites, even without overlapping, justifies the large number of satellites in constellation to be applied, to ensure the safe communication and the real-time continuity of services. Earth's surface area is 510 million km²; thus, the LEO satellite at altitude of 550 km under elevation of $40°$, covers the area of 0.00206×510 million km² $= 1.05$ million km², which is in fact the circled area on Earth with approximately radius of 580 km.

Table 5.3   Nadir angle and central angle for different elevations.

The horizon plane elevation (ε_0)	The first shell $H = 550$ km		The second shell $H = 1110$ km		The third shell $H = 340$ km	
	$\alpha_0(°)$	$\beta_0(°)$	$\alpha_0(°)$	$\beta_0(°)$	$\alpha_0(°)$	$\beta_0(°)$
Ideal: $0°$	66.9	23.1	58.3	31.7	71.6	18.4
Designed at: $25°$	56.4	8.6	50.4	14.6	59.3	5.7
Designed at: $30°$	52.8	7.2	47.5	12.5	55.2	4.8
Designed at: $35°$	48.9	6.1	44.2	10.8	51.0	4.0
Designed at: $40°$	44.8	5.2	40.7	9.3	46.6	3.4

Table 5.4 Coverage of the Starlink satellites.

The horizon plane elevation (ε_0)	The first shell H = 550 km C [%]	The second shell H = 1110 km C [%]	The third shell H = 340 km C [%]
Ideal: $0°$	4.003	7.461	2.55
Designed at: $25°$	0.560	1.614	0.247
Designed at: $30°$	0.394	1.185	0.175
Designed at: $35°$	0.283	0.885	0.121
Designed at: $40°$	0.206	0.657	0.088

5.7 Handover-Takeover Process: Geometrical Interpretation and Confirmation

To understand the global coverage and continuity of real-time services, the handover-takeover (known as just handover) process between two LEO satellites should be well understood, which is further geometrically interpreted and confirmed in its operation.

For tracking the satellites, real-time software fed by Kepler elements is applied. The respective software provides real-time tracking information, usually displayed in different modes. The "radar map" mode is further considered for the intended geometrical handover-takeover process interpretation. The "radar map" mode includes accurate satellite path with the ground station considered at the center, as in Figure 5.14 presented what is described in details under Section 4 (Cakaj, 2021).

For LEO satellites, the maximal elevation (*Max-El*) is the main parameter of the satellite pass over the ground station (user) and determines the communication duration between LEO satellite and the respective ground station. The horizon plane with a predetermined minimal elevation is considered the *designed horizon plane*. The designed horizon plane for users for the real-time uninterrupted communications is considered at 40° (usually ranges from 25° to 40°) (Starlink, 2020; Starlink Satellite Missions, 2020) and is presented as the thicker black circle indicated by 40° in the Figure 5.14.

For the geometrical interpretation and confirmation purposes are identified three randomly chosen orbits, as Orbit1, Orbit2, and Orbit3, and appropriate satellite to each of them at altitudes of 550 km. For each orbit is given an arrow identifying the satellite's movement direction seen from the user. The user is located at the center. At each orbit are identified points of satellite acquisition (AOS) and satellite loss (LOS) in space. Since this is only geometrical approach, the time as a variable is not considered.

The acquisition and loss of the satellites are considered for the ideal horizon plane at (0°) and for the designed horizon plane at (40°), designated as AOS (0), AOS (40), and LOS (0), AOS (40), respectively. These events in space are determined by respective azimuth and elevation. Each of three passes is determined by its own maximal elevation, and by the appropriate azimuth. For all points indicated in Figure 5.14, the coordinates are given in Table 5.5 as pairs of azimuth and elevation [Az °, El °]. The values are extracted from Figure 5.14, but approximated since the orbits are randomly chosen (Cakaj, 2021).

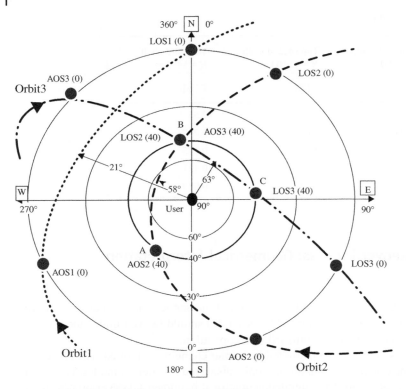

Figure 5.14 Geometrical interpretation of the handover-takeover process.

Table 5.5 Coordinates of the satellites space events.

Events for three orbits	Max-El [Az°, El°]	AOS (0) [Az°, El°]	AOS (40) [Az°, El°]	LOS (40) [Az°, El°]	LOS (0) [Az°, El°]
Orbit1	305°, 21°	240°, 0°	NA	NA	360°, 0°
Orbit2	310°, 58°	155°, 0°	220°, 40°	345°, 40°	30°, 0°
Orbit3	30°, 63°	315°, 0°	345°, 40°	85°, 40°	125°, 0°

Satellite flying at Orbit1, appears at the user's ideal horizon plane approximately at azimuth of 240° at 0° elevation, noted as event AOS1(0). The satellite moves higher, achieving the Max-El of 21° (at Az = 305°) and then down to the event LOS1(0) at coordinates [360°, 0°] and disappearing from the user's ideal horizon. The user and this satellite had no communication, since the satellite is always lower than user's designed horizon plane at 40° (21° < 40°), consequently not locked. The satellite in Orbit1, moves away unseen from the user. This fact explains NA (not applicable) of AOS (40) and LOS (40) for Orbit1 in Table 5.5.

Satellite flying at Orbit2, appears at the user's ideal horizon plane approximately at azimuth of 160° at 0° elevation, noted as the event AOS2(0) but not locked with the user. Satellite moves higher to the event AOS2(40) at coordinates (Az = 220°, 40°) where the satellite is locked with the user establishing communication. The range between the satellite and the user at this point is 809.5 km (apply Eq. 1.56). The satellite moves higher toward Max-El event at coordinates ($Az = 310°$, $58°$) still in communication with the satellite and being closest to it at the distance of 641.4 km (apply Eq. (1.56)). Satellite moves down, still in communication, toward the event LOS2(40) at coordinates ($Az = 345°$, $40°$) having again the range of 809.5 km. At this point the satellite leaves the user's designed horizon plane and being unlocked of the communication. Satellite moves further to the event LOS2(0) at coordinates ($Az = 30°$, $0°$) and leaving the user's ideal horizon plane also. For further purposes points AOS2(40) and LOS2(40) are, respectively, noted also as A and B. So, during the Orbit2, the user and the satellite had communication from point A to B, with range variation from 641.4 km to 809.5 km; consequently, one-way signal delay from 2.13 to 2.69 ms.

Satellite flying at Orbit3, appears at the user's ideal horizon plane approximately at azimuth of $315°$ at $0°$ elevation, noted as the event AOS3(0) but not locked with the user. The satellite moves higher to the event AOS3(40) at coordinates ($Az = 345°$, $40°$) where it is locked with the user establishing communication. The range between the satellite and the user at this point is 809.5 km (apply Eq. 1.56). The satellite moves higher toward Max-El event at coordinates ($Az = 30°$, $63°$) still in communication with the satellite and being closest to it at the distance of 611.2 km. The satellite moves down, still in communication, toward the event LOS3(40) at coordinates ($Az = 85°$, $40°$) once again at the range of 809.5 km. At this point, the satellite leaves the user's designed horizon plane and is unlocked – that is, it loses communication with the ground station. Satellite moves further to the event LOS3(0) at coordinates ($Az = 125°$, $0°$) and leaving the user's ideal horizon plane, also. For further purposes points AOS3(40) and LOS3(40) are, respectively, noted also as B and C. So, during the Orbit3, the user and the satellite had communication from point B to C, with range variation from 611.2 to 809.5 km – consequently one-way signal delay from 2.03 to 2.69 ms.

The user has communication with satellite of Orbit2 from point A to B. The same user has communication with satellite of Orbit3 from point B to C. Point B, identifies the loss of the satellite in Orbit2 from the user's designed horizon plane [event LOS2(40)] and the acquisition of the satellite in Orbit3 by the user's designed horizon plane [event AOS 3(40)]. If the satellite in Orbit2 can communicate by intersatellite communication protocol with the satellite in Orbit3 at this point (zone), establishing handover-takeover process between satellites in Orbit2 and Orbit3, will be ensured uninterrupted communication between the user and the constellation from point A to C. Thus, the handover process at point B ensures the continuity of services for the user, so keeping the communication from A to C. This geometrical interpretation confirms the continuity of services by two satellites. The same applies for more of them, for satellites that can intercommunicate within appropriate constellation (Cakaj, 2021).

Just one more clarification! Seems that the satellites hit each other at point B! No way! This is a very coordinated and synchronized process. To facilitate the handover-takeover process, the LEO satellites' coverage areas overlap by a few degrees. For this process, the satellites must be adjacent to each other and able to intercommunicate (Cakaj et al., 2010).

To further illustrate the handover-takeover process in space, let us assume that this process will happen at $39°$ elevation for satellite in Orbit2 (just $1°$ before leaving the user's horizon plane) and at $41°$ elevation for satellite in Orbit3 (just $1°$ before entering the user's horizon plane). The difference in $2°$ is ensured by overlapping. Can these two satellites communicate with each other to provide handover-takeover process between the satellite in Orbit2 and satellite in Orbit3? Seen from the user under $39°$, the distance between the satellite in Orbit2 and the ground station is 827.9 km (apply Eq. (1.56)), ready for handover and ready to leave the designed horizon plane. On the other side, the satellite in Orbit3 at elevation $41°$ is 800.6 km (apply Eq. (1.56)) far from the user, ready to enter into the user's designed horizon plane and to take over the communication from the satellite in Orbit2 by itself. In these positions, the satellites can communicate. Applying cosines rule, these satellites are far from each other around 40 km in space, so they easily can communicate with each other during the handover-takeover process. Finally, this is geometrical confirmation of the handover-takeover process, which proves the continuity of real-time services by satellites, including broadband worldwide internet services.

References

Alyildiz, I.F., Zunalioglu, H.M.D., and Bender, D.M. (1999). Handover management in low earth orbit (LEO) satellite networks. *Mobile Networks and Applications* 4 (1999): 301–310.

Cakaj, S. (2010). *Local User Terminals for Search and Rescue Satellites*, 84. Saarbrucken, Germany: VDM Publishing House.

Cakaj, S. (2016). *Coverage Belt for Low Earth Orbiting Satellites.*, 554–557. Opatija, Croatia: 39th International Convention on Information and Communication Technology, Electronics and Microelectronics, IEEE.

Cakaj, S. (2021). The parameters comparison of the "Starlink" LEO satellites constellation for different orbital shells. *Frontiers in Communications and Networks-Aerial and Space Networks* 2,, (Article ID: 643095): 1–15.

Cakaj, S., Fitzmaurice, M., Reich, J., and Foster, E. (2010). The downlink adjacent interference for low earth orbiting (LEO) search and rescue satellites. *International Journal of Communications, Networks and System Sciences (IJCNS)* 3 (2): 107–115.

Cakaj, S., Kamo, B., Lala, A., and Rakipi, A. (2014). The coverage analysis for low Earth orbiting satellites at low elevation. *International Journal of Advanced Computer Science and Applications* 5 (6): 6–10.

Cochetti, R. (2015). *Mobile Satellite Communications Handbook*, 119–155. Hoboken, NJ: Wiley Telecom.

De Selding, B.P. (2015) Virgin, Qualcomm invest in OneWeb satellite internet venture. SpaceNews (Jan 15).

Foust, J. (2018). Telesat to announce manufacturing plans for LEO constellation in coming months. *SpaceNews* (Feb. 18): https://spacenews.com/telesat-to-announce-manufacturing-plans-for-leo-constellation-in-coming-months/.

Jong, D.K., Goode, M., Liu, X. and Stone, M. (2014) Precise GNNS Positioning in Arctic Regions. *OTC Arctic Technology Conference*, Paper Number: OTC-24651-MS. Houston, Texas,

Papapetrou, P., Karapantazis, S., and Pavlidou, N.F. (2003). *Handover Policies in LEO Systems with Satellite Diversity.*, 10–11. Frascati, Italy: International Conference on Advanced Satellite Mobile Systems (ASMS).

Pultarova, T. and Henry, C. (2017). OneWeb weighing 2000 more satellites. SpaceNews, Feb2017.

Rao, J. (2019) How to see SpaceX's Starlink satellite 'train' in the night sky. https://www.space.com/spacex-starlink-satellites-night-sky-visibility-guide.html

Richharia, M. (1999). *Satellite communications systems*. New York: McGraw-Hill.

Roddy, D. (2006). *Satellite communications*. New York: McGraw Hill.

Seyedi, Y. and Safavi, M.S. (2012). On the analysis of random coverage time in Mobile LEO satellite communications. *Communications Letters, IEEE* 16 (5): 1–10.

Starlink (2020). https://en.wikipedia.org/wiki/Starlink#Global_broadband_Internet (Accessed January 17, 2021).

Starlink Satellite Missions (2020). https://directory.eoportal.org/web/eoportal satellite-missions/s/starlink (Accessed January 17, 2021).

Xiao, Y., Zhang, T., and Liu, L. (2018). Addressing subnet division based on geographical information for satellite-ground integrated network. *IEEE Access*, PP (99): 1–1. DOI:10.1109/ACCESS.2018.2882594. https://ieeexplore.ieee.org/document/8546744.

Zong, P. and Kohani, S. (2019). Optimal satellite LEO constellation design based on global coverage in one revisit. *International Journal of Aerospace Engineering*. https://doi.org/10.1155/2019/4373749.

6

LEOs Sun Synchronization

6.1 Orbital Sun Synchronization Concept

Earth is our home planet. The Earth rotates west to east, known as a *prograde direction*. Humans have known that the Earth is round for more than 2000 years! Even though our planet is a sphere, shown in Figure 6.1, it is not a perfect sphere (NASA. What is the Earth? 2020). Earth daily rotates at around 1000 mph (Gunn 2020). Earth's rotation, wobbly motion, and other forces are very slowly changing the planet's shape, flattening at the poles and bulging at the equator, but still keeping the roundish shape as an *oblate ellipsoid,* shown in Figure 6.2 (Knippers 2009).

An orbit in which one satellite moves in the same direction as the Earth's rotation is known as *prograde* or *direct orbit.* The inclination of a prograde orbit always lies between 0° and 90°. Most satellites are launched in a prograde orbit because the Earth's rotational velocity provides part of the orbital velocity, saving launch energy (Richharia 1999; Roddy 2006; Maral and Bousquet 2005). An orbit in which the satellite moves in direction opposite to that of the Earth's rotation is called *retrograde orbit.* The inclination of a retrograde orbit always lies between 90° and 180°. Sun synchronized orbits (SSOs), further discussed, are typically retrograde.

Microsatellites in low Earth orbits (LEOs) have been used for the last three decades, including for communications and scientific purposes. This chapter is dedicated mostly to scientific applications related to space observation and environmental monitoring.

Figure 6.3 gives an example of weather monitoring from satellites (ESA, Cyclone, west of Madagascar 2019). Such scientific missions are permanently developing, especially in fields where similar experiments by purely Earth-based means are now impracticable or too difficult to be implemented – including meteorology, oceanography, agriculture, biodiversity, forestry, landscape, geology, cartography, regional planning, education, intelligence, defense, and military.

These missions are mainly based on photo imagery records. The Sun's position is very important for photo imagery; thus, the observed area from the satellite to be treated under the same lighting (illumination) conditions. This can be achieved by keeping the orbital plane position constant relative to the Sun due to the Earth's motion around the Sun, what means that the satellite will observe a location on Earth at the same daily time, known as the Sun synchronization process, which is applicable only for LEO orbits.

Having these images under the same illumination conditions, of the different times, enables scientists or appropriate businesses to compare images and conclude about changes over time. Therefore, scientists use image series like these to investigate how weather patterns emerge, to help predict weather or storms; when monitoring emergencies like forest fires or flooding; or to

Ground Station Design and Analysis for LEO Satellites: Analytical, Experimental and Simulation Approach,
First Edition. Shkelzen Cakaj.
© 2023 The Institute of Electrical and Electronics Engineers, Inc. Published 2023 by John Wiley & Sons, Inc.

Figure 6.1 Earth rotating.

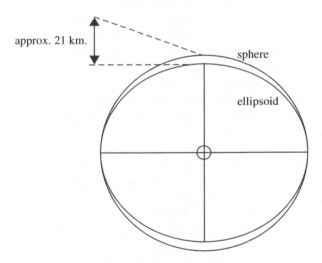

Figure 6.2 Earth's oblateness.

accumulate data on long-term problems like deforestation or rising sea levels (Polar and Sun synchronous orbit 2020). As better quality of the images the better scientific outcomes. On this trend, Pleiades Neo satellite optical constellation with four identical 30 cm resolution Sun synchronized satellites provides top-level Earth observation services now and going forward into decades to come (Pléiades Neo constellation 2021).

How are orbits synchronized with the Sun? The Earth is not a perfectly spherical homogeneous body. It is a round body characterized by a bulge at the equator, and a slight flattening at the poles.

Figure 6.3 Cyclone, west of Madagascar (ESA 2019).

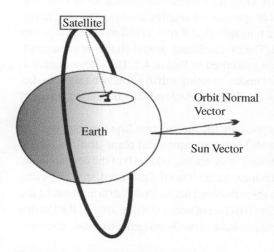

Figure 6.4 Orbital Sun synchronization concept.

Due to the irregularities of Earth's rotation and not homogenous Earth's mass distribution, the terrestrial potential at a point in space (in our case, the point indicates a satellite) depends not only on the distance r to the Earth's center but also on the respective longitude and latitude of the appropriate point within a spherical coordinate system. The terrestrial potential has been well studied by Kozai (1964) and then by Goddard Space Flight Center, which developed "Goddard Earth Models," denoted as GEM-1, GEM-2, GEM-3, and GEM-4 (Geopotential model 2022, Maral and Bousquet 2005). For both the Kozai and Goddard approaches, it is common to express the terrestrial potential in terms of geopotential zonal harmonic coefficients J_n. The J_2 term due to flattening of the Earth (about 21 km, Figure 6.2) dominates all other harmonics. Based on Kozai (1964) and Maral and Bousquet (2005), the J_2 harmonic is nondimensional value given as:

$$J_2 = 1.0827 \cdot 10^{-3} \tag{6.1}$$

For missions accomplished based on photo imagery, in order to keep and treat the observed area under the same illumination conditions due to the different satellite paths over the observed area on the Earth, somehow the orbit (orbital plane) should be managed (controlled) to keep the same position related to the Sun. Geometrically, this could be interpreted as keeping a constant angularity between the orbital plane and the Sun, or better defined, keeping constant angularity between the perpendicular (normal) vector on the orbital plane and the Sun direction vector, as given in Figure 6.4 (Cakaj et al. 2009). Since the Earth's rotation around the Sun takes one year, for

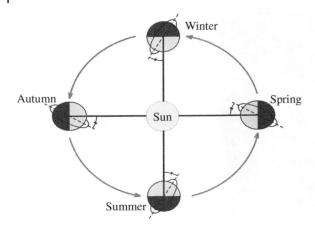

Figure 6.5 Seasons Sun synchronization.

consistency purposes, the orbit should maintain the appropriate position relative to Sun throughout the various seasons of the year, as presented in Figure 6.5 (Richharia 1999; Roddy 2006; Maral and Bousquet 2005). Orbits that fulfill this condition are known as Sun Synchronized Orbits (SSOs).

To a first approximation, the orbital plane is fixed in space as the satellite orbits around the Earth. But, due to the different potentials generated by the non-spherical Earth, at different orbital points in space, variations of the orbital elements happen. The most affected orbital elements are *ascending/descending nodes* and the *argument of perigee*, as presented in Figure 4.2. The affected ascending/descending nodes consequently affect the line of nodes, creating a drift of the orbital plane. Let us first discuss the effect on *ascending/descending nodes (line of nodes)* and later on the argument of perigee deviation.

From Figure 4.2, we see that the line of nodes represents the intersection line (interconnecting ascending node, Earth's center, and descending node) between equatorial plane and the orbital plane. Line of nodes is determined by the right ascension of ascending node (Ω) of the orbital plane, as in Figure 4.2. If the orbital plane drifts because of nonuniform terrestrial potential, so will the line of nodes, manifested as the slide of line of nodes on the equatorial plane always drifting around the Earth's center, as the central point of the line of nodes. This line of nodes shifting around the Earth's center on the equatorial plane, because of Earth's oblateness and nonhomogeneous mass, is known as the *nodal regression*.

6.2 Orbital Nodal Regression

The line of nodes determines the orientation of the satellite's orbital plane in space. Nodal regression refers to the shift of the orbit's line of nodes over time as Earth revolves around the Sun, due to the Earth's oblate nature. Nodal regression is an especially useful feature for the low circular orbits providing to them the Sun synchronization. A Sun-synchronized orbit lies in a plane that maintains a fixed angle with respect to the Earth–Sun direction. Nodal regression depends on orbital altitude and orbital inclination angle. In the low Earth, Sun synchronized circular orbits are suited satellites that accomplish their photo imagery missions. If the orbit is prograde, the line of nodes slides westward, and if it is retrograde, it slides eastward. This means that nodes (line of nodes) move in opposite direction to the direction of satellite motion, hence the term *nodal regression*.

For the LEO, determined by semimajor axis a and its eccentricity e laid on the orbital plane in space, which is determined by inclination i and the right ascension of the ascending node Ω, an approximate expression for the nodal rate regression of Ω due to time is expressed as (Maral and Bousquet 2005):

$$\frac{d\Omega}{dt} = -\left(\frac{3}{2}\right) n_0 A J_2 \cos i \tag{6.2}$$

where J_2 is the second harmonic of the terrestrial geopotential, and n_0 is mean satellite movement:

$$n_0 = \frac{2\pi}{T} \tag{6.3}$$

where T is orbital period. Furthermore, A is constant given as:

$$A = \frac{R_E^2}{a^2(1-e^2)^2} \tag{6.4}$$

$R_E = 6378$ km is Earth's radius, e is orbital eccentricity, i is the inclination, and a is a semimajor axis of satellite's orbit.

For circular orbits, approximately $e = 0$ and $a = r$, where r is orbital radius of circular orbit. Orbital period for circular orbits is expressed as:

$$T = 2\pi\sqrt{\frac{r^3}{\mu}} \tag{6.5}$$

where $\mu = 3.986005 \cdot 10^5$ km^3/s^2 is Earth's geocentric gravitational constant. For circular orbit yields out:

$$A = \frac{R_E^2}{r^2} \tag{6.6}$$

Substituting R_E, μ and J_2, finally results in nodal regression expressed by inclination i and orbital radius r. The nodal regression expressed in (°/day) is (Cakaj et al. 2013).

$$\Delta\Omega = -2.06474 \cdot 10^{14} \cdot \frac{\cos i}{r^{7/2}} \ [°/\text{day}] \tag{6.7}$$

From Eq. (6.7), the nodal regression for circular orbits depends on orbit inclination and orbital altitude (radius). The nodal regression becomes zero in the case of the inclination angle being 90°. When the orbit inclination angle is $i < 90°$, then deviation is negative, so according to Eq. (6.2) the satellite orbital plane rotates in a direction opposite to the direction of the Earth's rotation.

When the orbit inclination angle is $i > 90°$, then deviation is positive, so the satellite orbital plane rotates in the same direction as the direction of the Earth's rotation. From the above statements we see that if the orbit is prograde (eastward), the nodes slide westward, and if it is retrograde (westward), the nodes slide eastward. Thus, nodes (line of nodes) move in the opposite direction to the direction of satellite motion, hence the term *nodal regression*.

Idea: Considering the Eq. (6.2) or Eq. (6.7), the idea is to draw conclusions about the range of nodal regression under different altitudes and inclination.

Method: Consider Van Allen belt effect (Van Allen radiation belt 2020) for simulation purposes at attitudes from 600 up to 1200 km. LEOs have very low eccentricity, which can be considered $e \approx 0$ and then $a = r$. Thus, for altitudes from 600 up to 1200 km and considering Earth's radius as $R_E \approx 6400$ km, the orbits' radius range is 7000 to 7600 km.

Results: Considering Eq. (6.7) for these orbits and assumed radius ranges, the nodal regression [°/day] for different inclination angles i are calculated and presented in Table 6.1 and Figure 6.6. Figure 6.6, shows that for the eastward regression, inclination i must be greater than 90°; consequently, the orbit must be retrograde. This feature enables an orbital plane to stay in constant

Table 6.1 Nodal regression (°/day).

Inclination [°]	Orbital radius [km]			
	7000	7200	7400	7600
20	−6.74	−6.12	−5.55	−5.06
30	−6.21	−5.64	−5.12	−4.66
40	−5.50	−4.99	−4.52	−4.13
50	−4.61	−4.19	−3.78	−3.46
60	−3.59	−3.26	−2.95	−2.69
70	−2.45	−2.22	−2.02	−1.84
80	−1.24	−1.13	−1.02	−0.94
90	0	0	0	0
100	1.24	1.13	1.02	0.94
110	2.45	2.22	2.02	1.84
120	3.59	3.26	2.95	2.69
130	4.61	4.19	3.78	3.46
140	5.50	4.99	4.52	4.13
150	6.21	5.64	5.12	4.66
160	6.74	6.12	5.55	5.06

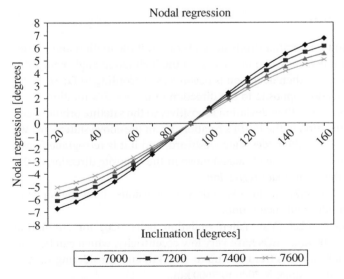

Figure 6.6 Nodal regression [°/day].

angular relationship with the Sun throughout the seasons, which will be proved by the next section. The nodal regression is inversely dependent on $r^{7/2}$. Considering the LEO satellite altitude of 1200 km, the nodal regression is around 6.7°. Considering MEO altitudes of 10 000 km, around 10 times more than LEO altitudes, applying the Eq. (6.7), the nodal regression is almost negligible. This means that medium or geosynchronous orbits cannot be Sun synchronized.

Conclusion: These results confirm that nodal regression for attitudes from 600 up to 1200 km may range from 0° up to 6.7°, depending on inclination angle. Lower inclination causes higher deviation. For inclination of 90°, there is no nodal deviation. Only LEOs can be Sun synchronized; thus, the photo imagery mission can be accomplished only by LEO satellites, to provide high performance.

6.3 LEO Sun Synchronization and Inclination Window

Section 6.2 defined the correlation between line of nodes deviation ($d\Omega$) over time and inclination. The sign ($-$) is because of the opposite direction of the satellite's motion and the nodal deviation (nodal regression). This is the feature, applied for Sun-synchronization. But how?

An orbital plane fixed with respect to the Earth effectively makes a 360° rotation in space in a year (about 365.25 days), since Earth itself rotates by 360° around the Sun. This rate is equivalent to a rotation of the orbital plane of about 0.986 [°/day] (Cakaj et al. 2009). By choosing a pair of particular values of i and r, it is possible to obtain an orbit for which the nodal regression varies each day by a quantity equal to the rotation of the Earth around the Sun. Mathematically, this is expressed as:

$$\frac{d\Omega}{dt} = 0.9856°/day \tag{6.8}$$

Applying Eq. (6.8) at (6.7):

$$\Delta\Omega = -2.06474 \cdot 10^{14} \cdot \frac{cos\ i}{r^{7/2}} = 0.9856\ [°/day]. \tag{6.9}$$

From this equation it follows that for eastward regression, inclination i must be greater than 90°; thus, the orbit should be retrograde. Figure 6.7 shows the SSO, which obviously has an inclination i greater than 90° (Sunsynchronous orbit 2019). A satellite in a Sun-synchronous orbit would usually be at an altitude of between 600 and 800 km.

Under the condition expressed by Eq. (6.8), the angle between perpendicular vector on the orbital plane and the Sun direction vector remains constant throughout the year, or having the constant angularity during the year. This feature is known as LEO Sun synchronization. It is confirmed through the previous section that only LEOs can be Sun synchronized.

Idea: The further question is, what is the inclination window for LEO Sun synchronization for the range of altitudes for LEO orbits?

Method: Mathematics and simulation are applied. By solving the Eq. (6.8), for orbital altitude of 600 km consequently for $a = r = 7000$ km under no eccentricity ($e = 0$) will get inclination for Sun synchronization as:

$$i_1 = 97.9 \tag{6.10}$$

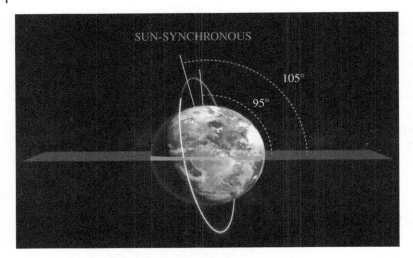

Figure 6.7 Sun synchronized orbit (SSO).

and for orbital attitude of 1200 km consequently for $a = r = 7600$ km under no eccentricity ($e = 0$) will get inclination for sun synchronization as:

$$i_2 = 100.5°. \tag{6.11}$$

Further, considering these values of inclination, but also the range for lower and higher orbital attitudes (600–1200 km), the nodal regression for the inclination from 97° up to 101° (Sun-synchronized inclination window) is calculated.

Results: These results are given in Table 6.2 and Figure 6.8 (Cakaj et al. 2013).

In Figure 6.8 the nodal regression of 0.986[°/day] in fact represents the daily angular rotation of the Earth around the Sun.

Conclusion: Sun-synchronization for LEO circular orbits depend on altitude and inclination. SSOs are retrograde. The orbital Sun synchronization is achieved through inclination angle at range of 97.9–100.5° for LEO altitudes from 600 to 1200 km. The inclination window for Sun synchronization has a width of 2.6° for altitudes of 600 to 1200 km. A Sun-synchronization feature is typical for LEOs. At medium and high orbits (MEO or GEO), the nodal regression effect is negligible, so these orbits could not be Sun synchronized.

Table 6.2 Nodal regression for sun-synchronized inclination window (°/day).

Inclination [°]	Orbital radius [km]			
	7000	7200	7400	7600
97	0.876	0.793	0.721	0.656
98	1.001	0.906	0.824	0.750
99	1.125	1.018	0.926	0.843
100	1.248	1.131	1.028	0.936
101	1.372	1.242	1.129	1.028

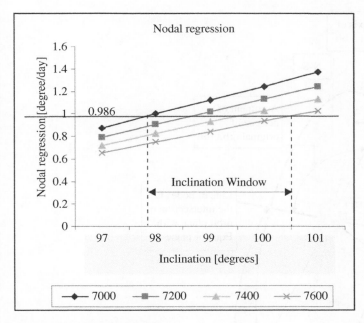

Figure 6.8 Inclination window.

6.4 Perigee Deviation under Inclination Window for Sun-Synchronized LEOs

The orientation of the orbit within orbital plane is defined by the *argument of perigee ω* (Figure 4.2) The argument of the perigee determines the position of the major axis. This is the angle, taken positively from 0° to 360° in the direction of the satellite's motion, between the direction of the line of nodes and the direction of the orbital perigee. Like the line of nodes shifting due to equatorial bulge of Earth, the argument of perigee undergoes natural perturbation, also. This is defined as *orbital perigee deviation*.

Idea: The further idea is to calculate this orbital perigee deviation for Sun-synchronized orbits, or better, saying how much the major axis deviates for one satellite pass for Sun-synchronized orbits. This perturbation may be visualized as the movement of the elliptical orbit in a fixed orbital plane, as presented in Figure 6.9 (Figure presents the case of only perigee deviation under unchangeable line of nodes position). This deviation happens in a fixed orbital plane (two orbits in Figure 6.9 lie in the same plane; it is in fact an original and a deviated orbit). Obviously, the apogee and perigee points change in position, manifested as a major axis drift, deviated by $\Delta\omega$. The question is, how much?

Method: Mathematics and simulation are applied. For analytical and simulation purposes, the altitudes from 600 to 1200 km are considered. Further, to determine perigee deviation under inclination window of the Sun-synchronized orbits, we simulate the orbital perigee deviation for the considered altitudes and the eventual impact on the satellite's mission. This deviation is a function of the satellite altitude and orbital plane inclination angle. This drift is analyzed under all serious

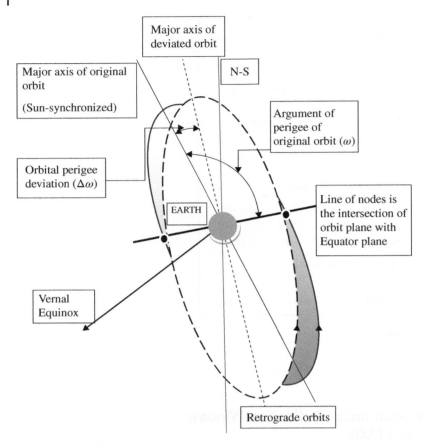

Figure 6.9 Orbital perigee deviation.

satellite books, and for simulation purposes is applied here (Maini and Agrawal 2011, p. 83), expressing the perigee deviation per orbit in [°/orbit] as:

$$\Delta\omega = 0.29 \left[\frac{4 - 5\sin^2 i}{(1 - e^2)^2} \right] \left[\left(\frac{D}{r_a + r_p} \right)^2 \right] \tag{6.12}$$

This is the general expression for elliptical orbit, where i is inclination, e is orbit eccentricity, r_a, r_p are orbital apogee and perigee, and D is Earth's diameter. Based on Eq. (6.12), there is no perigee deviation under the inclination of 63.4° (the feature is applied for Molnya orbit) since:

$$4 - 5\sin^2(63.4°) = 0 \rightarrow \Delta\omega = 0 \tag{6.13}$$

For the inclination lower than 63.4°, then, $\Delta\omega$ is positive, so the perigee deviation occurs in the same direction as the satellite motion, and for the inclination greater than 63.4°, $\Delta\omega$ is negative, so the perigee deviation occurs in the opposite direction as the satellite's motion. Also, the closer the satellite is to the center of the Earth, the larger is the deviation.

For simulation purposes, consider altitudes from 600 to 1200 km, and Earth's radius as $R_E \approx$ 6400 km; then the orbits' radius ranges from 7000 to 7600 km. From the previous section, based on Eq. (6.9), we know that the range of orbital inclination lies from 97.9° to 100.5° to attain Sun synchronization (Cakaj et al. 2013). The inclination within this range is always greater than 63.4°; thus, the result of Eq. (6.12) is always negative, which means that the orbital major axis will drift in the opposite direction to the satellite's motion.

LEO are usually circular orbits, so the eccentricity is 0. Also, $r_a = r_p = a$, where a is a semi-major axis and $D = 2R_E$, where $R_E \approx 6400$ km is Earth's radius. Applying these statements at Eq. (6.12), yields the perigee deviation for low Earth Sun-synchronized orbits.

$$\Delta\omega = 0.29 \left(\frac{R_E}{a} \right) \left[4 - 5 \sin^2 i \right] \tag{6.14}$$

For Sun-synchronization, the orbital inclination lies on range:

$$97.9° \leq i \leq 100.5° \tag{6.15}$$

Finally, the orbital perigee deviation for any Sun-synchronized orbit lies in the range of:

$$\Delta\omega_{(i=97.9)} \leq \Delta\omega \leq \Delta\omega_{(i=100.5)} \tag{6.16}$$

Results: The calculation of orbital perigee deviation and appropriate results are presented in Table 6.3 (Cakaj and Kamo 2019).

The negative sign indicates that the orbital perigee deviation shifts in opposite side to the satellite motion. From the Table 6.3, it is obvious that the largest perigee deviation appears at altitude of 7000 km at inclination of 97.9° and the lowest perigee deviation appears at altitude of 7600 km at inclination of 100.5°. These values converted in minutes, represent the deviation of 13.1′ and 10.3′ per orbit, respectively, for the largest and lowest case.

Finally, the orbital perigee deviation expressed in [°/orbit], for LEO Sun-synchronized orbits is mathematically expressed as Cakaj and Kamo (2019):

$$10.3' \leq \Delta\omega \leq 13.1' \quad [°/orbit] \tag{6.17}$$

This calculation leads toward the thrust vector being applied in order to keep argument of perigee under a predefined value, respectively unchangeable over time. The vector intensity depends on absolute value of perigee deviation.

Conclusion: Sun-synchronized orbits are very useful for Earth's observation (scientific satellites) applications. Nodal regression is especially utilized for LEO circular orbits, providing them Sun synchronization. SSOs are always retrograde because the inclination is greater than 90°. Due to Earth's equatorial bulge and natural perturbations, the argument of the perigee also deviates. This is known as orbital perigee deviation, expressed in °/orbit. For the considered LEO altitudes under the Sun synchronized inclination window, the orbital perigee deviation ranges from 10.3 to 13.1 (′/orbit), always in the opposite direction of the satellite motion.

Table 6.3 Perigee deviation for Sun-synchronized orbits.

Inclination [°]	Orbit radius [km]			
	7000	7200	7400	7600
97.9	−0.219	−0.207	−0.196	−0.186
100.5	−0.202	−0.191	−0.180	−0.171

References

Cakaj, S. and Kamo, B. (2019). Orbital perigee deviation under inclination window for Sun synchronized low Earth orbits. *EJERS, European Journal of Engineering Research and Science* 4 (10) Brussels: 127–130.

Cakaj, S., Fischer, M., and Schotlz, L.A. (2009). Sun Synchronization of Low Earth Orbits (LEO) through Inclination Angle. In: *28th IASTED International Conference on Modelling, Identification and Control, MIC 2009*, Innsbruck, Austria, 155–161.

Cakaj, S., Kamo, B., and Malaric, M. (2013). The inclination window for low Earth sun synchronized satellite Orbits. *Transactions on Maritime Science* 2 (1) Split, Croatia: 15–19.

ESA, Cyclone, west of Madagascar, (2019). https://www.esa.int/ESA_Multimedia/Images/2019/03/Cyclone_Idai_west_of_Madagascar

Geopotential model (2022). https://en.wikipedia.org/wiki/Geopotential_model

Gunn, A. (2020). What is the speed rotation! *Science Focus* https://www.sciencefocus.com/planet-earth/earth-rotation-speed.

Knippers, R. (2009). *Geometric Aspects of Mapping*. Enschede, The Nederland: International Institute for Geo-Information Science and Earth Observation (ITC).

Kozai, Y. (1964). New determination of zonal harmonics coefficients of the earth's gravitational potential. *Astronomical Society of Japan* 16: 263. Bibliographic Code: 1964PASJ...16...263K.

Maini, K.A. and Agrawal, V. (2011). *Satellite Technology*, 2e. West Sussex: Wiley.

Maral, G. and Bousquet, M. (2005). *Satellite Communications Systems*, 4e. West Sussex: Wiley.

NASA (2020). What is the Earth? https://www.nasa.gov/audience/forstudents/5-8/features/nasa-knows/what-is-earth-58.html

Pléiades Neo constellation (2021). https://www.intelligence-airbusds.com/imagery/constellation/pleiades-neo/?gclid=EAIaIQobChMI1_2_tdPZ9QIV1uFRCh0lRQIhEAAYASAAEgIzJ_D_BwE

Sunsynchronous orbit (2019) https://uzayteknesi.com/2019/11/30/yorunge-tipleri-sun-synchronous-orbit/

Richharia, M. (1999). *Satellite Communications Systems*. New York: McGraw Hill.

Roddy, D. (2006). *Satellite Communications*, 4e. New York: McGraw Hill.

Van Allen radiation belt. https://en.wikipedia.org/wiki/Van_Allen_radiation_belt.

7

Launching Process

7.1 Introduction to the Launching Process

An orbit is the trajectory followed by the satellite. The communication between the satellite and a ground station is established when the satellite is consolidated in its own orbit. Different types of orbits are possible, each suitable for a specific application or mission. The most used orbits are circular – categorized as low, medium, and geosynchronous (geostationary) orbits based on the altitude above the Earth's surface. Communications satellites are often placed in the geostationary orbit, since the best coverage is achieved. Almost the whole Earth can be covered with three geostationary satellites on the equatorial plane, in angular equidistance of 120°, as shown in Figure 7.1. The geostationary orbit is a circular orbit 35 786 km above the Earth's Equator and follows the direction of the Earth's rotation.

Rockets are used to place the satellite in the appropriate orbit. The main goal of the rocket launch is to enable the satellite to acquire the desired space orbital parameters. Planning the flight path of a rocket is very complicated, since the gravitational pull works against the rocket; thus, the process is organized in several phases and as a structure consists of several subsystems (Lucket 2013).

In addition to the interest for satellites' operation and missions, people are often interested in studying the flight path of the rocket sending the satellite in space, with special attention on trying to record the appropriate flight. The ISS (International Space Station) flight in January 2012 is recorded over Florida, under perfect night sky viewing conditions. I was surprised with the clarity of the night sky photo, so I included it within this book, presented in Figure 7.2 (Night Sky, 2012).

The most difficult orbit to be attained, generally, is a geosynchronous Earth orbit (GEO) since they are at the highest altitude above the Earth's surface (a geostationary orbit, also abbreviated GEO, is a special case of the geosynchronous orbit, in which there is no inclination). Moreover, for the geostationary orbit, the inclination must be brought on zero degree to be aligned with the equatorial plane.

Because GEO satellites are such a large distance from the Earth's surface, placing the satellite in geostationary orbit is almost impossible to attain in one step. There are two methods for putting satellites in a geostationary orbit:

1) The three-step process is when a rocket takes the spacecraft to a low Earth circular orbit, then toward geostationary transfer orbit, and finally into the circular geostationary orbit.
2) The low circular orbit is skipped and the satellite goes straight to geostationary transfer orbit, and finally into the circular geostationary orbit.

Ground Station Design and Analysis for LEO Satellites: Analytical, Experimental and Simulation Approach, First Edition. Shkelzen Cakaj.
© 2023 The Institute of Electrical and Electronics Engineers, Inc. Published 2023 by John Wiley & Sons, Inc.

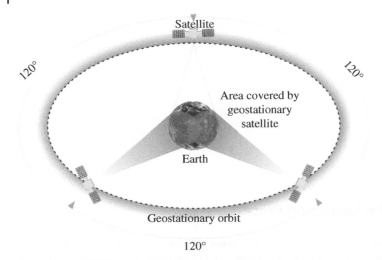

Figure 7.1 Three geostationary satellites equal-spaced by 120°.

Figure 7.2 International Space Station Flight Over Florida (Night Sky, 2012).

The three-step process is known as *Hohmann transfer*. In the first step, the satellite is injected into a circular low Earth orbit (LEO). In the second step, the satellite's orbit is transformed from the LEO into an elliptical transfer orbit (also known as a parking orbit) by maneuvers at perigee, in order to attain the apogee equal to geostationary (GEO) orbit's radius. Finally, the satellite is moved from the elliptical transfer orbit to the final destination, as a geostationary orbit (Maini and Agrawal 2010; Lorenzini et al. 2000), as shown in Figure 7.3 (Hohmann transfer orbit, 2020).

The Hohmann transfer orbit is based on two instantaneous velocity changes. The transfer consists of a velocity impulse on an initial circular orbit, in the direction of motion and collinear with velocity vector, which propels the space vehicle into an elliptical transfer orbit. The second velocity impulse also in the direction of motion is applied at apogee of the transfer orbit, which propels the space vehicle into a final circular orbit at the final geostationary altitude (Arluklar et al. 2012; Kamel 2011).

Figure 7.3 Hohmann transfer.

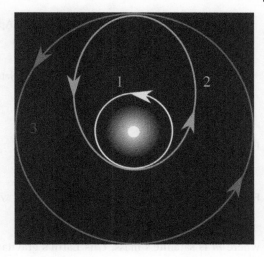

The Hohmann transfer, which involves two circular orbits with different orbital inclinations, is known as non-coplanar Hohmann transfer. If both orbital planes, from low orbit to transfer orbit and from the transfer to the geostationary plane, are aligned, then the Hohmann transfer is known as coplanar, discussed in detail later in this chapter. In terms of propellant consumptions, the Hohmann transfer is most common method for transferring between circular coplanar orbits.

Further in this section, we will analyze the involvement of LEOs in the process of attaining the geostationary orbit. Analyses are approached step by step, first addressing the creation of the transfer elliptical orbit from the LEO, simulating the thrust vector to be applied at the perigee to attain the desired transfer orbit. Further, the Hohmann transfer is simulated under different LEOs altitudes. Finally, the section looks at the launching process from a Kourou, French Guiana, launching site in detail.

From the aerospace dynamics perspective, there are two challenges to make the satellite operational:

1) The energy needed (velocity) at the orbit injection point for the satellite to enter and follow the orbit controlled by the Earth's gravity.
2) The velocity of the satellite at each orbital point.

The first condition speaks to the satellite's correct and safe orbital acquisition and the second to the satellite's consolidation in its orbital "bed," which is always controlled by Earth's gravity!

The body will have a "peaceful" movement in its orbit if it behaves according to the first law of thermodynamics, known as the *law of conservation of energy,* which states that the energy of a closed system must remain constant – it can neither increase nor decrease without outside interference. Figure 7.4 shows the Earth and the satellite as a closed system without being disturbed from the outside. Applying the law of energy conversion, it is:

$$E_{P} + E_{K} = CONSTANT = -G\frac{M \cdot m}{2a} \tag{7.1}$$

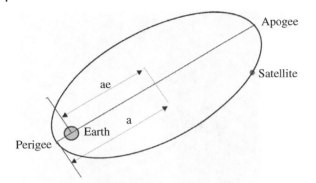

Figure 7.4 The Earth and the satellite as a closed system.

where m is satellite's mass, G is Earth's gravitational constant, and M is Earth's mass, denoted together as $\mu = GM = 3.986 \cdot 10^{14}(m^3/s^2)$. Furthermore, expressions for both kinetic and potential energy are given as:

$$E_K = \frac{1}{2}mv^2 \text{ and } E_P = -G\frac{Mm}{r} \tag{7.2}$$

where r is random distance between satellite and the Earth's center. Applying Eq. (7.2) at Eq. (7.1) yields:

$$v = \sqrt{\mu\left(\frac{2}{r} - \frac{1}{a}\right)} \tag{7.3}$$

representing the velocity of the satellite distanced by r from the Earth center, at any point in the elliptical orbit with semimajor axis a.

The transfer orbit due to the Hohmann transfer is the elliptical orbit; thus, let us clarify some geometrical features of the elliptical orbit. The elliptical orbit is determined by the *semi-major axis*, which defines the size of an orbit, and the *eccentricity* that defines the orbit's shape. Orbits with no eccentricity are known as *circular orbits*. The elliptical orbit is shaped as an ellipse, where the satellite is farthest from the Earth's center at the *apogee* (r_a) and closest at the *perigee* (r_p), presented in Figure 7.5.

The orbit's eccentricity is defined as the ratio of difference to sum of apogee (r_a) and perigee (r_p) radii as Eq. (7.4) (Richharia 1999; Roddy 2006):

$$e = \frac{r_a - r_p}{r_a + r_p} \tag{7.4}$$

Applying geometrical features of ellipse yield the relations between semi-major axis, apogee, and perigee:

$$r_p = a(1 - e) \tag{7.5}$$

$$r_a = a(1 + e) \tag{7.6}$$

$$2a = r_a + r_p \tag{7.7}$$

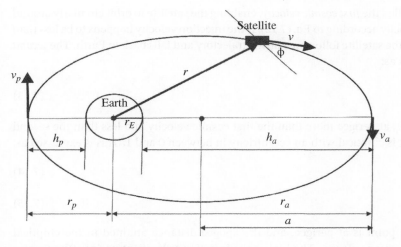

Figure 7.5 Parameters of an elliptical orbit.

both r_p and r_a are considered from the Earth's center. Earth's radius is $R_E = 6371$ km. Then, the altitudes (highs) of perigee and apogee are:

$$H_p = r_p - R_E \tag{7.8}$$

$$H_a = r_a - R_E \tag{7.9}$$

Thus, permanent altitude control is mandatory. Different algorithms are applied and active control means are generally added to assure accurate altitude stabilization, keeping the altitude errors within permitted limits – consequently, keeping the satellite stability within its orbit and the quality of the communication (Reyhanoglu and Drakunov 2008; Esmailzadeh et al. 2011).

7.2 Injection Velocity and Apogee Simulation from Low Earth Orbits

The best way to describe the orbit implementation process is in terms of cosmic velocities. Based on Kepler's laws, considering an elliptical orbit, the satellite's velocity at the perigee and apogee point, based on Eq. (7.3), respectively, are expressed as (Maini and Agrawal 2010):

$$v_p = \sqrt{\left(\frac{2\mu}{r_p}\right) - \left(\frac{2\mu}{r_a + r_p}\right)} \tag{7.10}$$

$$v_a = \sqrt{\left(\frac{2\mu}{r_a}\right) - \left(\frac{2\mu}{r_a + r_p}\right)} \tag{7.11}$$

$\mu = GM = 3.986 \cdot 10^{14} (m^3/s^2)$. For orbit with no eccentricity ($e = 0$), apogee and perigee distances are equal ($r_a = r_p = a$).Thus, the orbit becomes circular with radius a and orbital velocity as:

$$v_1 = \sqrt{\frac{\mu}{a}} \tag{7.12}$$

By definition, this is called the *first cosmic velocity*, enabling the satellite to orbit circularly around the Earth at uniform velocity according to Eq. (7.12). If the injection velocity happens to be less than the first cosmic velocity, the satellite follows a ballistic trajectory and falls back to Earth. The *second cosmic velocity* is defined as:

$$v_2 = \sqrt{\frac{2\mu}{a}} \tag{7.13}$$

For injection velocity v_p at perigee more than the first cosmic velocity and less than the second cosmic velocity, the orbit is elliptical with an eccentricity in between 0 and 1. This is expressed as:

$$v_1 < v_p < v_2 \tag{7.14}$$

$$0 < e < 1 \tag{7.15}$$

The satellite injection point is at perigee, and the apogee distance attained in the elliptical orbit depends on the injection velocity. The higher the injection velocity at perigee, the greater is the apogee distance, as schematically presented in Figure 7.6. For the same perigee distance, the radius of perigee, r_p, if under the injection velocity v_{p1} at perigee it attains an apogee distance, radius of apogee, r_{a1}, and under velocity v_{p2} it attains an apogee distance r_{a2}, then applying Eq. (7.10) the relationship between velocities at perigee and respective attained distances at apogee is:

$$\left(\frac{v_{p2}}{v_{p1}}\right)^2 = \frac{1 + \dfrac{r_p}{r_{a1}}}{1 + \dfrac{r_p}{r_{a2}}} \tag{7.16}$$

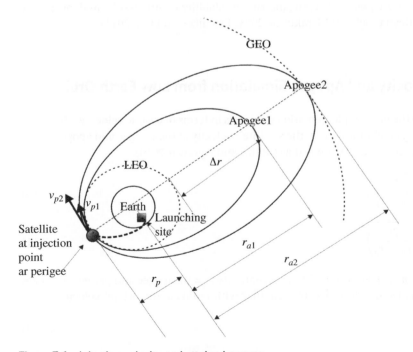

Figure 7.6 Injection velocity and attained apogee.

By definition, the apogee distance is always larger than the perigee distance; thus, it can be expressed as:

$$r_a = r_p + \Delta r \tag{7.17}$$

where Δr represents the distance of how much it is intended to achieve larger apogee compared with perigee of the orbit. This is defined as *apogee incremental step* (Cakaj et al. 2012).

Idea: What perigee velocity should be applied to attain the apogee larger than the perigee for an advanced defined incremental apogee step (Δr)? This is, in fact, related to the second Hohmann launching procedural step, from the LEO to the transfer (parking) orbit (In Figure 7.3, passing from 1 to 2).

Method: Math and simulation are applied. Let us use Eq. (7.17); applying Eq. (7.4) and (7.10), will lead to the eccentricity and the perigee velocity as:

$$e = \frac{\Delta r}{\Delta r + 2r_p} \tag{7.18}$$

$$v_p = \sqrt{\frac{2\mu}{r_p} \cdot \frac{\left(1 + \dfrac{\Delta r}{r_p}\right)}{\left(2 + \dfrac{\Delta r}{r_p}\right)}} \tag{7.19}$$

Eq. (7.18) expresses how the eccentricity changes with Δr, respectively, and how the eccentricity changes with the apogee incremental step, keeping the fixed perigee. Eq. (7.19) tells us which injection velocity v_p has to be applied at perigee point in order to attain the apogee for Δr larger than a predefined perigee. For $\Delta r = 0$, the orbit is circular ($e = 0$), and according to Eq. (7.19) orbital velocity is the first cosmic velocity [$v_1 = \sqrt{\mu/r}$].

For simulation purposes, the fixed perigee is considered. Four fixed perigee distances are considered as: 7000, 7200, 7400, and 7600 km. Considering the Earth's radius, $R_E \cong 6400$ km, these perigee distances correspond to the altitudes above the Earth's surface approximately of 600, 800, 1000, and 1200 km, which are very common altitudes for low Earth orbits, at the first step of the Hohmann launching process.

The apogee increase is considered by n steps, as (Cakaj et al. 2012):

$$n \cdot \Delta r = n \cdot 2000 \text{ km} \tag{7.20}$$

$n = 0,1,2,3,\dots$

For $n = 0$, the orbit is circular with low altitude, and for $n = 18$ the apogee distance achieves approximately the radius of geostationary orbit. In between these two cases fall medium altitudes for medium orbit satellites, if Hohmann transfer is applied to attain medium Earth orbits (MEO). For different values of n the eccentricity variations and the perigee velocity are calculated to attain the appropriate apogee.

Results: Based on this approach, firstly the eccentricity variation is analyzed because of apogee increment, presented in Figure 7.7, which confirms the eccentricity increment with apogee increase. Finally, the goal behind the idea is to draw conclusions about the required velocity at injection perigee point as a function of the apogee increment. For this purpose, we apply Eq. (7.19). The results are presented in Table 7.1 and Figure 7.8.

Figure 7.8 shows four curves for different perigee values, respectively, corresponding to LEO altitudes of 600, 800, 1000, and 1200 km. From, Table 7.1, for altitude of 600 km (perigee of 7000 km),

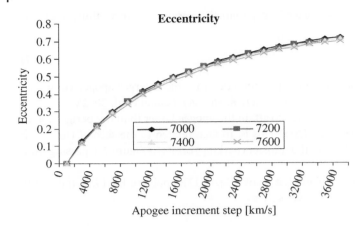

Figure 7.7 Eccentricity variation.

Table 7.1 Injection velocity for intended apogee increment [km/s].

Apogee Increment Step Δr [km]	Perigee highs (r_p) [km]			
	7000	7200	7400	7600
0	7.55	7.44	7.34	7.24
2000	8.00	7.88	7.76	7.65
4000	8.34	8.21	8.08	7.96
6000	8.60	8.46	8.33	8.20
8000	8.81	8.67	8.53	8.40
10 000	8.98	8.83	8.69	8.56
12 000	9.12	8.97	8.83	8.69
14 000	9.24	9.09	8.95	8.81
16 000	9.34	9.19	9.05	8.91
18 000	9.43	9.28	9.13	8.99
20 000	9.51	9.36	9.21	9.07
22 000	9.58	9.42	9.28	9.14
24 000	9.64	9.48	9.34	9.20
26 000	9.69	9.54	9.39	9.25
28 000	9.74	9.59	9.44	9.30
30 000	9.79	9.63	9.48	9.34
32 000	9.83	9.67	9.52	9.38
34 000	9.86	9.71	9.56	9.42
36 000	9.90	9.74	9.59	9.45

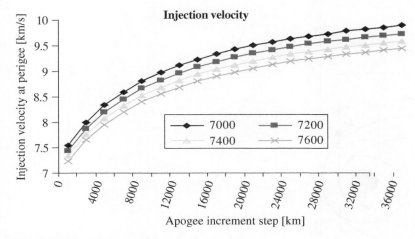

Figure 7.8 Injection velocity and attained apogee.

the injection velocity at perigee point is 7.55 km/s, and for altitude of 1200 km (perigee of 7600 km) the injection velocity at perigee is 7.24 km/s. At higher perigee (higher altitude), the lower injection velocity is required at perigee point (Cakaj et al. 2012). Obviously, for defined LEO, the higher velocity is required at perigee in order to attain larger apogee. We can use Figure 7.8 to determine the required velocity at perigee point to attain respective apogee.

Let us consider this relation from the point of view of different LEO altitudes. For example, in order to attain an apogee of 42 400 km (corresponding to geostationary altitude) from the LEO with a perigee of 7400 km (low earth altitude of 1000 km), the applied velocity at perigee must be 9.59 km/s. In order to attain the same apogee from the higher LEO of altitude of 1200 km (perigee of 7600 km), the lower velocity must be applied at perigee point, concretely as 9.45 km/s. This means that in order to attain the geostationary altitude, less velocity must be applied at a higher LEO altitude. If less energy is invested to bring the satellite at low Earth orbit, then more energy is need for the geostationary transfer orbit.

Conclusion: The process of passing from the low orbit toward transfer elliptical orbit is discussed as the second step of the Hohmann transfer. For the fixed perigee of the elliptical transfer orbit, the attained apogee depends on injection velocity at perigee point. The greater injection velocity from the first cosmic velocity, the greater is the apogee distance attained. Curves provided can be applied to find out the attained apogee height for a given value of injection velocity at the perigee point, or on the other hand for required apogee what injection velocity has to be applied. It is confirmed that in order to attain apogees of (7000–42 400) km the injection velocity to be applied at perigee point ranges at (7.24–9.90) km/s.

7.3 Hohmann Coplanar Transfer from Low Earth Orbits

In orbital mechanics, the Hohmann transfer orbit is an elliptical orbit used to transfer between two circular orbits of different radii. If both orbits lie in the same plane, it is known as coplanar transfer (Curtis 2005; Kamel 2011). (Intentionally, first is discussed the coplanar transfer, to be clarified, since the following section takes care further about nonplanarity).

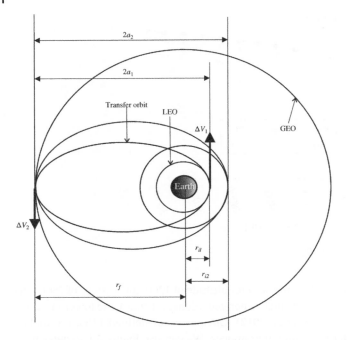

Figure 7.9 Hohmann coplanar transfer orbit.

The first (low radius) circular orbit is defined as *initial* (r_{in}) and the second (high radius) is defined as a *final* orbit (r_f). The orbital maneuver to perform the Hohmann transfer applies two engine impulses (thrusts), one to move a spacecraft onto the transfer orbit and a second to move it to final orbit. For the coplanar Hohmann transfer, two applied velocity impulses are confined to the orbital planes of the initial and final orbits. The transfer from the low radius orbit to the high radius orbit is attained in three steps. The first one is the launch of the satellite in the LEO. This is the initial circular orbit with radius of (r_{in}) as it is presented in Figure 7.9. By the second step, the first velocity impulse (Δv_1) is applied at the low Earth orbit creating an elliptic transfer orbit with perigee altitude equal of the initial circular orbit and the apogee altitude equal to the final circular orbit. Finally, at the third step, the second velocity impulse (Δv_2) is applied at apogee of the transfer orbit in order to attain the final orbit, respectively the GEO with radius (r_f) as presented in Figure 7.9. The apogee of the transfer orbit is equal to the radius of the final orbit. Thus, the second velocity impulse circularizes the transfer orbit at apogee. Both velocity impulses (Δv_1 and Δv_2) keep the direction of the orbits' motion and are confined to the orbital planes of the initial and final orbits, as seen in Figure 7.9 (Cakaj et al. 2015).

In Figure 7.9, r_f represents the radius of geosynchronous orbit. Since our goal is to calculate the thrust (impulse) velocities to be applied at LEO orbits of different radiuses, thus r_{i1}, $r_{i2}...r_{in}$ denote the radiuses of initial low Earth circular orbits (i – indicates just initial). This fact causes different major axis of transfer orbits as $2a_1$, $2a_2$, ...$2a_n$.

Under these considerations, r_f remains unchangeable and it is the radius of geosynchronous orbit as $r_f = 42\ 164$ km. Thus, for major axis stems out the relation:

$$2a_n = r_f + r_{in} \tag{7.21}$$

$n = 1, 2, ...N$, where N indicates number of initial low Earth orbits.

Based on Eq. (7.12) will have the velocity of initial circular orbit as v_{in} and the velocity of final circular orbit as v_f. Thus, v_{in} depends on LEO radius r_{in} and v_f remains unchangeable because of r_f unchangeability, expressed as follows:

$$v_{in} = \sqrt{\frac{\mu}{r_{in}}} \tag{7.22}$$

$$v_f = \sqrt{\frac{\mu}{r_f}} \tag{7.23}$$

Idea: The transfer is initiated by firing the space craft engine at LEO in order to accelerate it, so that it will follow the elliptical orbit; this adds energy to the spacecraft's orbit. When the spacecraft has reached transfer orbit, its orbital speed (and hence its orbital energy) must be increased again in order to change the elliptic orbit to the larger circular one, respectively, to a geosynchronous (geostationary) one. Thus, the further idea is to calculate these velocity pulses for the different LEO altitudes (H_{in}) (Cakaj et al. 2015).

Method: Mathematics and simulations are applied. For elliptic orbit with perigee equal to r_{in} and apogee of r_f, the velocities at perigee and apogee are:

$$v_{pn} = \sqrt{\left[\left(\frac{2\mu}{r_{in}}\right) - \left(\frac{2\mu}{r_{in} + r_f}\right)\right]} \tag{7.24}$$

$$v_{an} = \sqrt{\left[\left(\frac{2\mu}{r_f}\right) - \left(\frac{2\mu}{r_{in} + r_f}\right)\right]} \tag{7.25}$$

v_{pn} is in fact the velocity in the transfer orbit at initial orbit height and v_{an} in fact is the velocity in the transfer orbit at final orbit height.

The initial velocity increment (Δv_{1n}), to move the satellite from the initial circular orbit to elliptic transfer orbit, is given as the difference between velocity at perigee of transfer orbit v_{pn} and the velocity of the initial circular orbit v_{in}:

$$\Delta v_{1n} = v_{pn} - v_{in} \tag{7.26}$$

The final velocity increment (Δv_{2n}), to move the satellite from elliptic transfer orbit to geosynchronous circular orbit, is given as the difference between the velocity on the final circular v_f orbit and the velocity on the apogee of the transfer elliptical orbit v_{an}:

$$\Delta v_{2n} = v_f - v_{an} \tag{7.27}$$

Applying (7.24) to (7.26) and (7.25) to (7.27), will stem out:

$$\Delta v_{in} = \sqrt{\frac{\mu}{r_{in}}}\left(\sqrt{\frac{2r_f}{r_{in} + r_f}} - 1\right) \tag{7.28}$$

$$\Delta v_{2n} = \sqrt{\frac{\mu}{r_f}}\left(1 - \sqrt{\frac{2r_f}{r_{in} + r_f}}\right) \tag{7.29}$$

Further, it defined normalized radius R_n as follows:

$$R_n = \sqrt{\frac{r_{in}}{r_f}} \tag{7.30}$$

from which one yields out:

$$v_f = R_n \cdot v_{in} \tag{7.31}$$

Applying Eqns. (7.30) and (7.31) at (7.28) and (7.29), finally will have velocity increments to be applied at Hohmann coplanar transfer orbit in order to attain geosynchronous orbit from different low Earth orbits, as:

$$\Delta v_{1n} = v_{in} \cdot \left(\sqrt{\frac{2}{R_n^2 + 1}} - 1 \right) \tag{7.32}$$

$$\Delta v_{2n} = v_{in} \cdot R_n \cdot \left(1 - R_n \cdot \sqrt{\frac{2}{R_n^2 + 1}} \right) \tag{7.33}$$

From the propellant consumption point of view (Pearson 1999), it is of interest the contribution of both velocity impulses as:

$$\Delta v_n = \Delta v_{1n} + \Delta v_{2n} \tag{7.34}$$

The eccentricity of transfer orbit for different low Earth orbits is given as:

$$e_n = \frac{1 - R_n^2}{1 + R_n^2} \tag{7.35}$$

By the end, just to remind that: $H_{in} = r_{in} - R_E$, where $R_E = 6371$ km is Earth's radius.

For simulation purposes five altitudes of LEOs are considered as initial orbits for the Hohmann transfer, starting from altitude of 500 km up to 1300 km, which are typical for LEO satellites. For each of them it is calculated (Δv_1) and (Δv_2). Also, (Δv) is calculated (Cakaj et al. 2015).

Results: These results are presented in Table 7.2 and in Figure 7.10.

Conclusion: It is confirmed that the higher the altitude of the low initial orbit, the lower velocity impulse is needed to the final destination orbit. Consequently, less fuel is needed to be carried out on the space craft in case that the space craft initially is injected on the higher altitude. For different altitudes of initial orbit, the first velocity impulse has faster decreased gradient than the second velocity impulse.

Table 7.2 Velocity impulses to be applied for Hohmann coplanar transfer.

n	1	2	3	4	5
H_{in} (km)	500	700	900	1100	1300
r_{in} (km)	6878	7078	7278	7478	7678
v_{in} (km/s)	7.587	7.479	7.376	7.277	7.181
R_{in}	0.403	0.409	0.415	0.421	0.426
e_n	0.721	0.713	0.706	0.699	0.692
Δv_{1n} (km/s)	2.359	2.311	2.257	2.204	2.161
Δv_{2n} (km/s)	1.443	1.422	1.405	1.384	1.364
Δv (km/s)	3.802	3.733	3.662	3.588	3.525

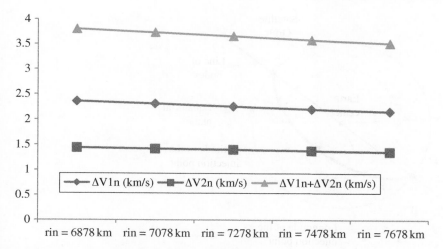

Figure 7.10 Velocity variations under different initial orbit altitudes.

7.4 The GEO Altitude Attainment and Inclination Alignment

Usually, to attain the geosynchronous or geostationary Earth orbit from LEO, the propellant makes up a considering part of a global satellites mass, what directly impacts the launching process and satellite's lifetime. When launching and then consolidating the satellite on its own high circular orbit the needed satellite's propellant mass must be minimized.

The Hohmann transfer is well known for the minimum of propellant mass used for satellite transfer into high orbits. The Hohmann transfer orbit is based on two instantaneous velocity changes. The transfer consists of a velocity impulse on an initial circular orbit, in the direction of motion and collinear with velocity vector, which propels the space vehicle into an elliptical transfer orbit. The second velocity impulse also in the direction of motion is applied at apogee of the transfer orbit, which propels the space vehicle into a final circular orbit at the final altitude (Bonavito et al. 1998; Cakaj et al. 2015).

After too long efforts and experiments, in January 2013, the Robotic Refueling Mission (RRM) performed its first test on board the ISS (International Space station), demonstrating that remotely controlled robots successfully transfer fuel in space. The fuel transport in space, to inject more propellant in the satellite, for the continuity of services, creates two advantages: the lower launching cost and more room for instruments to be packed aboard. In the near future, it will be possible to robotically refuel satellites in orbits, allowing satellite service providers to dramatically extend their services and the satellites' lifetime (Robotic Refueling Mission 2016).

The launching process toward geostationary orbit from the site in Kourou, French Guiana, is analyzed, associated with maneuvers, and thrust to be applied so the geostationary orbit can be attained and consolidated. Maneuvers and applied thrusts are directly correlated with propellant consumption/saving (Figure 7.10).

The main goal of the launching process is to enable the satellite to acquire the desired space orbital parameters, given in Figure 4.2. In order to attain the exact desired orbit plane, it is vital that correct conditions are established at the satellite injection point. Further, to stay in the desired orbit position to carry out the satellite's mission, various in orbit operations such as orbit stabilization, orbit correction, and stations keeping are mandatory.

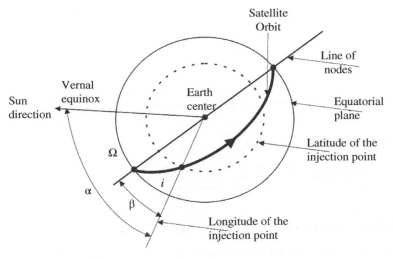

Figure 7.11 Space orbit parameters and injection point.

The angle defining the right ascension of the ascending node Ω, is basically the difference between two angles, α and β, where α is the angle made by the longitude of the injection point at the time of launch with the direction of vernal equinox and β is the angle made by the longitude of the injection point at the time of launch with the line of nodes, as shown in Figure 7.11, and given as (Maini and Agrawal, 2010):

$$\Omega = \alpha - \beta \tag{7.36}$$

Considering Figure 7.11, it is obvious that to ensure the satellite orbits within a given plane, the satellite must be injected at a specific time, depending on the longitude of the injection point, at which the line of nodes makes the required angle with the direction of the vernal equinox. Thus, for a known angle of inclination and the longitude of injection point (launching site), the launching time must be exactly determined in order to acquire the desired ascension of the ascending node Ω (Maini and Agrawal, 2010, Roddy 2006).

The angle of inclination i of the orbital plane can be determined from the known values of the azimuth angle (A_Z) and the latitude (l) of the injection point, expressed as (Maini and Agrawal, 2010):

$$cosi = sin\,A_Z cosl \tag{7.37}$$

The azimuth angle (A_Z) at a given point in a satellite orbit is the angle made by the projection on the local horizon of the satellite vector velocity at that point with the north.

From Eq. (7.37), for the angle of inclination to be zero (for geostationary), the right-hand side of (7.37) should be 1, which could happen only if the launching site is at the Equator; thus, the latitude $l = 0°$ and $A_Z = 90°$ ($sin\,A_Z = 1$). Launching sites above the equatorial plane, it is $l > 0$ and $sin\,A_Z < 1$, which will decrease $cosi$ and consequently increase the inclination angle i, mathematically expressed as (Maini and Agrawal, 2010):

$$i > l \tag{7.38}$$

Thus, we can conclude that the satellite will tend to orbit in a plane, which will be inclined to the equatorial plane at the angle greater than the latitude of the injection point.

Figure 7.12 Satellite launch site at Kourou (Guiana Space Center Kourou, (2021)).

Further, we discuss the satellite launch process from the launch site Kourou in French Guiana toward final geostationary orbit. The location of Kourou at French Guiana was selected in 1964 to become the spaceport of France and it is operational since 1968. In 1975, France offered to share Kourou with ESA (European Space Agency) (Europe's Space Port 2016). The satellite launch site at Kourou in French Guiana is located at coordinates of 5° 9′ 34.92″ N, 52° 39′ 1.08″ W, and it is particularly suitable as a location for a spaceport, as it fulfills the two major geographical requirements of such a site. It is quite close to the Equator, and it has uninhabited territory, so that lower stages of rockets and debris from launch failures cannot fall on human habitations. The aerial view of the vector launch area of the Kourou site on October 22, 2021, in Kourou, French Guiana, presented in Figure 7.12.

The entire satellite launching process by Ariane vehicle toward geostationary orbit, launched from Kourou is further described by activity steps. The orbit circularization and the geostationary are achieved through two phases, the first one is the circular altitude attainment at 35786 km (geosynchronous orbit altitude) and the second is the inclination alignment (zero inclination) with Earth's equatorial plane.

7.4.1 Circularization and the Altitude Attainment

The launch vehicle takes the satellite to a point that is intended to be the perigee of the transfer orbit, at a height of about 200 km above the Earth's surface. At this point, the satellite is ejected from the rocket's nose cone (see Figure 7.12, conic nose on top), and it is injected directly in the transfer orbit (skipping the low orbit). The satellite, along with its apogee boost motor, is injected before the launch vehicle crosses the equatorial plane.

The injection velocity at perigee v_p expressed as:

$$v_p = \sqrt{\left(\frac{2\mu}{r_p}\right) - \left(\frac{2\mu}{r_a + r_p}\right)} \tag{7.39}$$

is such that the injected satellite attains an eccentric elliptical orbit with an apogee altitude at about 35 786 km where the velocity is v_a expressed by:

$$v_a = \sqrt{\left(\frac{2\mu}{r_a}\right) - \left(\frac{2\mu}{r_a + r_p}\right)} \tag{7.40}$$

r_p and r_a are, respectively, perigee and apogee distances (from the Earth's center), $\mu = GM = 3.986 \cdot 10^{14}(m^3/s^2)$, G is the Earth's gravitational constant and m Earth's mass. The correlation in between apogee (r_a), perigee (r_p) distances with respective altitudes at apogee (H_a) and at perigee (H_p) is:

$$r_{a,p} = R_E + H_{a,p} \tag{7.41}$$

For launching conditions from the site at Kourou toward geostationary orbit, and considering Earth's radius $R_E = 6371$ km, we find that the perigee and apogee distances, respectively, are $r_p = 6571$ km and $r_a = 42\ 159$ km (Cakaj et al. 2016). Applying these parameters in (7.39) and (7.40) reveals that the velocity at the injection point (perigee) in order to create a geostationary transfer orbit is:

$$v_p = 9.248 \text{ km/s} \tag{7.42}$$

and the velocity at apogee of the geostationary transfer orbit is:

$$v_a = 1.589 \text{ km/s} \tag{7.43}$$

The main goal is to place the satellite in the proper orbit, but usually it is not all done at once. For this purpose, the Apogee Kick Motor (AKM) is usually used. The case of three separate burns generating three velocity thrusts is presented in Figure 7.13.

Considering that the initial altitude at the injection point is at 200 km, the first burn (thrust) strives to attain an altitude of about 12 000 km, the second one an altitude of about 24 000 km, and finally the third thrust pushes the satellite to the geostationary altitude of 35 786 km, further applying Eq. (7.39), (7.40), and (7.41); the respective velocities and altitudes attained up to the final orbit circularization are given in Table 7.3.

Table 7.3 confirms that under the third thrust at apogee point, the geostationary altitude is attained and the orbit is circularized. The variation of velocities after each thrust is presented in Figure 7.14, which shows the convergence of both velocities at apogee and perigee after the third thrust, when the orbit is stabilized and consequently the satellite has constant velocity at each point of the orbit. In this case, where only the orbit circularization altitude attainment is needed, to increase the perigee distance, the thrust velocity vectors at apogee point are coplanar at each burn (thrust), and the thrust is applied in the same direction of the satellite's motion. Each velocity vector is tangential at apogee point of the orbit (see Figure 7.13).

The satellite is always moving with an actual velocity at apogee; thus, in order to increase the perigee, the increment on apogee velocity vector by the apogee thrust must be applied, expressed as:

$$\Delta v_{a(i)} = v_{a(i)} - v_{a(i-1)} \tag{7.44}$$

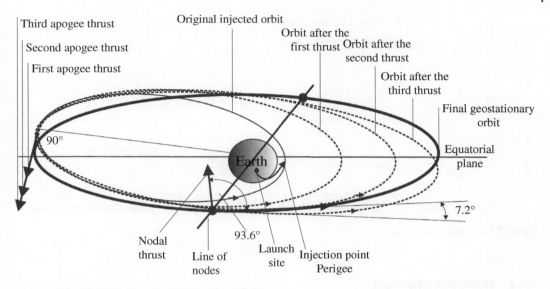

Figure 7.13 Three thrusts at apogee and nodal thrust.

Table 7.3 Velocities and respective altitudes (Cakaj et al. 2016).

	Velocity at perigee (v_p) [km/s]	Altitude at perigee (H_p) [km]	Velocity at apogee (v_a) [km/s]	Altitude at apogee (H_a) [km]
Injection phase	9.248	200	1.589	35 786
Under thrust 1	5.501	12 000	2.383	35 786
Under thrust 2	3.910	24 000	2.801	35 786
Under thrust 3 (Final stage)	3.067	35 786	3.067	35 786

where i indicates burning-thrust step and $i-1=0$ indicates injection phase. Finally, based on Table 7.3 and Eq. (7.44) the intensities of coplanar thrust vectors applied three times at apogee for orbit circularization are:

$$\Delta v_{a1} = 794[\text{m/s}] \qquad \Delta v_{a2} = 418[\text{m/s}] \qquad \Delta v_{a3} = 266[\text{m/s}] \qquad (7.45)$$

This orbit circularization is achieved by three appropriate thrust maneuvers without affecting any change of the inclination. For the attained circularization and altitude of 35 786 km, the satellite velocity in the circularized orbit is

$$v_a = 3.067\,[\text{km/s}] \qquad (7.46)$$

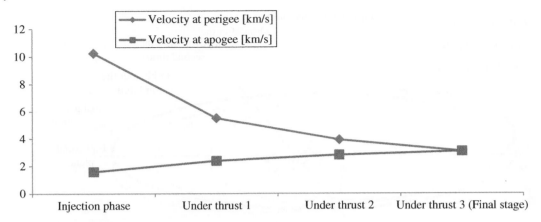

Figure 7.14 Velocities variation.

7.4.2 Inclination Alignment

The projection of the injection velocity vector in the local horizon plane at launching site of Kourou makes an azimuth angle of 85°. The latitude of the launch site is 5° 9′ 34.92″. The inclination angle attained by the geostationary transfer orbit is calculated from (7.37), and is:

$$i = 7.2^{\circ} \tag{7.47}$$

Next, another thrust is applied to bring the orbit inclination to 0°, respectively, to align the already circularized orbit with the equatorial plane, and finally the geostationary is to be completed.

The change in inclination Δi can be externally affected by applying a thrust velocity vector Δv_i at angle of $(90° + \Delta i/2)$ at one of nodes, as illustrated in Figure 7.13. The thrust is given by (Maini and Agrawal, 2010):

$$\Delta v_i = 2v\sin\left(\frac{\Delta i}{2}\right) \tag{7.48}$$

where v is the satellite's velocity at circularized orbit at Eq. (7.46) and Δi is the inclination range that should be corrected. Since the already circularized orbit has inclination of 7.2°, which should be brought at 0°, then $\Delta i/2$ is 3.6°. Finally, the velocity thrust to be applied for orbit alignment with equatorial plane, and making the orbit fully geostationary, is:

$$\Delta v_i = 385.15[m/s] \tag{7.49}$$

This thrust is applied at either of the two nodes (ascending or descending), as in Figure 7.13. The last step is fine tuning for the consolidation of the satellite at the appropriate space slot within the geostationary orbit.

Conclusion: It is confirmed that the geostationary orbit is achieved by three thrusts applied at apogee and a single thrust applied at the nodal point. Three apogee thrusts will achieve geostationary orbit altitude and circularity, and then the nodal thrust will align the satellite orbit with Earth's equatorial plane. Apogee thrust vectors are tangential with orbit and the nodal vector is perpendicular to the orbit plane.

References

Arluklar, V.P., Naik, S.D., and D.S. (2012). Optimality of generalized Hohmann transfer unsing dynamical approach lamber solution. *International Journal of Applied Mathematics and Mechanics* 8 (10): 28–37.

Bonavito, L.N., Der, J.G., and Vinti, P.J. (1998). *Orbital and Celestial Mechanics*. ARC (Aerospace Research Central).

Cakaj, S., Kamo, B., and Agastra, E. (2016). The altitude attainment and inclination alignment for the satellite launched from Kourou site. *Transactions on Networks and Communications* 4 (2): 39–46.

Cakaj, S., Kamo, B., Kolici, V., and Shurdi, O. (2012). The injection velocity and apogee simulation for transfer elliptical satellite orbits. *International Journal of Communications, Network and System, Sciences (IJCNS)* 5 (3): 187–191.

Cakaj, S., Kamo, B., Lala, A. et al. (2015). The velocity increment for hohmann coplanar transfer from different low Earth orbits. *Frontiers in Aerospace Engineering* 4 (1) Destech Publications, Inc,: Lancaster, Pennsylvania.

Curtis, H. (2005). *Orbital Mechanics for Engineering Students*. Oxford: Elsevier Butterworth-Heinemann http://www.nssc.ac.cn/wxzygx/weixin/201607/P020160718380095698873.pdf.

Esmailzadeh, R., Arefkhani, H., and Davoodi, S. (2011). Active control and attitude stabilization of a momentum-biased satellite without yaw measurements. In: *19th Iranian Conference on Electrical Engineering (ICEE)*, 1–6.

Europe's Space Port (2016). http://www.esa.int/Our_Activities/Launchers/Europe_s_Spaceport/Europe_s_Spaceport2

Guiana Space Center Kourou (2021). https://www.gettyimages.in/detail/news-photo/aerial-view-of-the-vector-launch-area-of-the-kourou-news-photo/1360699210?adppopup=true

Hohmann transfer orbit (2020). https://en.wikipedia.org/wiki/Hohmann_transfer_orbit#/media/File:Hohmann_transfer_orbit2.svg

Kamel, M.O. (2011). The generalized Hoohmann transfer with plane change using energy concepts. *Mechanics and Mechanical Engineering*. 15 (2): 183–191.

Lorenzini, C.E., Cosmo, L.M., Kaiser, M. et al. (2000). Mission analysis of spinning Systems for Transfers from low orbits to geostationary. *Journal of Spacecraft and Rockets* 37 (2): 165–172.

Lucket, E. (2013). *Digital Detection of Rocket Apogee*. Houston, Texas: Rice University.

Maini, A. and Agrawal, K. (2010). *Satellite Technology*. UK: Wiley.

Night Sky (2012). https://stock.adobe.com/in/images/the-night-sky/228754215

Pearson, J. (1999). Low cost launch systems and orbital fuel depot. *Acta Astronautica* 18 (4): 315–320.

Reyhanoglu, M. and Drakunov, S. (2008). Attitude Stabilization of Small Satellites Using Only Magnetic Actuation. In: *34th Conference on Industrial Electronics (IECON)*, 103–107.

Richharia, M. (1999, 1999). *Satellite communication systems*. New York: McGraw Hill.

Robotic Refueling Mission (2016) http://ssco.gsfc.nasa.gov/rrm_refueling_task.html [2016, February].

Roddy, D. (2006). *Satellite communications*. New York: McGraw Hill.

8

LEO Satellites for Search and Rescue Services

8.1 Introduction to LEO Satellites for Search and Rescue Services

Low Earth orbit (LEO) satellites are used for public communications, for scientific purposes, and also for humanitarian missions such as search and rescue services, described in this chapter. NOAA (National Oceanic and Atmospheric Administration) LEO environmental satellites provide continuous coverage of Earth, supplying high-resolution global meteorological, oceanic, and space observation data. In addition, these satellites are part of the international COSPAS – SARSAT system (COsmicheskaya Systyema Poiska Aariynyich Sudov – Search and Rescue Satellite Aided Tracking), designed to provide distress alert and location data in order to assist on search and rescue operations. Ground stations for such purposes are established in various locations throughout the world, as shown in Figure 8.1 (SRC: COSPAS – SARSAT system 2012). Around 37,000 persons worldwide have been rescued since 1982 because of SARSAT (International COSPAS-SARSAT program 2013).

COSPAS-SARSAT is an international, humanitarian satellite-based search and rescue system that operates continuously, detecting and locating transmissions from emergency beacons carried by ships, aircrafts, and individuals. This system was originally sponsored by Canada, France, the former Soviet Union, and the USA (NOAA-SARSAT 2022; COSPAS-SARSAT. Int 2020). The success of the rescue operation – whether aircraft, marine, or individual distress – depends on accurate rapid determination of the distress location. And once located, it is vital to communicate that location to rescue authorities. This depends on the communication reliability between the LUTs (local user terminals) and the satellites (Ludwig et al. 1985; Bulloch 1987) The basic COSPAS – SARSAT concept is illustrated in Figure 8.2 (NOAA-SARSAT 2022).

1) In situations anywhere in the world when lives are at risk, the emergency beacons are activated manually or automatically (Number [1] in Figure 8.2).
2) Emergency alerts received by the satellites are retransmitted (Number [2] in Figure 8.2), to ground stations established worldwide, with several more being built each year. These satellite ground stations are so called LUTs, (Number [3] in Figure 8.2).
3) Alerts are routed to a Mission Control Center (MCC) (Number [4] in Figure 8.2) in the country that operates LUT. Routed alerts include beacon location computed at the LUT received by one of the system LEO satellites.
4) After validation processing (based on the Doppler effect), alerts are relayed, depending on beacon location or country of registration (beacons of 406 MHz) to either another MCC or to an

Ground Station Design and Analysis for LEO Satellites: Analytical, Experimental and Simulation Approach,
First Edition. Shkelzen Cakaj.
© 2023 The Institute of Electrical and Electronics Engineers, Inc. Published 2023 by John Wiley & Sons, Inc.

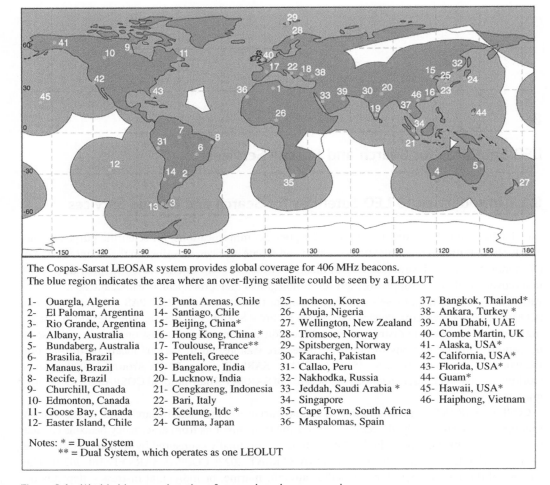

The Cospas-Sarsat LEOSAR system provides global coverage for 406 MHz beacons.
The blue region indicates the area where an over-flying satellite could be seen by a LEOLUT

1- Ouargla, Algeria	13- Punta Arenas, Chile	25- Incheon, Korea	37- Bangkok, Thailand*
2- El Palomar, Argentina	14- Santiago, Chile	26- Abuja, Nigeria	38- Ankara, Turkey *
3- Rio Grande, Argentina	15- Beijing, China*	27- Wellington, New Zealand	39- Abu Dhabi, UAE
4- Albany, Australia	16- Hong Kong, China *	28- Tromsoe, Norway	40- Combe Martin, UK
5- Bundaberg, Australia	17- Toulouse, France**	29- Spitsbergen, Norway	41- Alaska, USA*
6- Brasilia, Brazil	18- Penteli, Greece	30- Karachi, Pakistan	42- California, USA*
7- Manaus, Brazil	19- Bangalore, India	31- Callao, Peru	43- Florida, USA*
8- Recife, Brazil	20- Lucknow, India	32- Nakhodka, Russia	44- Guam*
9- Churchill, Canada	21- Cengkareng, Indonesia	33- Jeddah, Saudi Arabia *	45- Hawaii, USA*
10- Edmonton, Canada	22- Bari, Italy	34- Singapore	46- Haiphong, Vietnam
11- Goose Bay, Canada	23- Keelung, ltdc *	35- Cape Town, South Africa	
12- Easter Island, Chile	24- Gunma, Japan	36- Maspalomas, Spain	

Notes: * = Dual System
       ** = Dual System, which operates as one LEOLUT

Figure 8.1 Worldwide ground stations for search and rescue services.

appropriate Rescue Coordination Center (RCC) (Number [5] in Figure 8.2) (NOAA-SARSAT 2022; COSPAS-SARSAT. Int 2020).

The SARSAT refers to the USA segment, which is further considered for analysis. The author spent several months at NOAA as a postdoc researcher analyzing the operation of LUTs in terms of simulation, operation, and performance, from which study stems this chapter.

8.2 SARSAT System

US portion of COSPAS-SARSAT system is operated by the NOAA SARSAT Office in Suitland, Maryland. NOAA environmental satellites carry SARSAT packages. The United States Mission Control Center (USMCC) is also located in Suitland. US RCCs are operated by the US Coast Guard and the US Air Force (NOAA-SARSAT 2022; COSPAS-SARSAT. Int 2020).

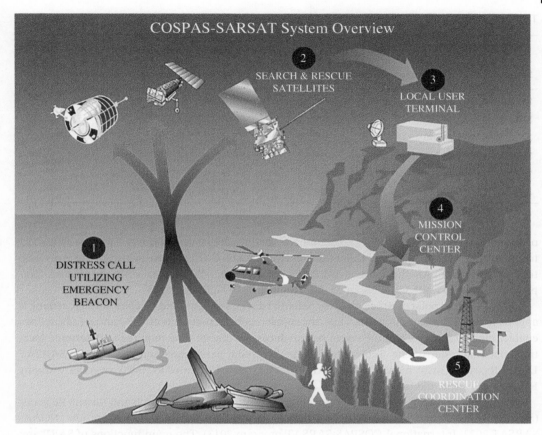

Figure 8.2 COPSAS - SARSAT concept.

8.2.1 SARSAT Space Segment

The space segment of the SARSAT system uses two types of satellites: polar-orbiting satellites in low Earth orbits and satellites in geosynchronous orbit (GEO), presented in Figure 8.3 (NOAA-SARSAT 2022; COSPAS-SARSAT. Int 2020).

LEO and GEO satellites complement each other on search and rescue services. GEO satellites continually view large areas of the Earth, and can provide immediate alerting and identification

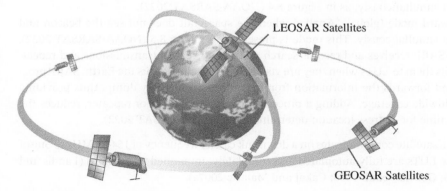

Figure 8.3 GEO and LEO satellites.

Table 8.1 SARSAT LEOSAR orbital parameters and operational status.

Satellite	Orbital parameters and payload instruments				
	Mean motion (rev/day)	Altitude (km)	Orbit period (hh:mm:ss)	406 MHz	Global
SARSAT-7	14.2475	809.45	01:41:04.2	F	F
SARSAT-8	14.1251	850.91	01:41:56.7	F	F
SARSAT-9	14.2405	811.80	01:41:07.2	F	F
SARSAT-10	14.1125	855.21	01:42:02.2	F	F
SARSAT-11	14.2149	820.43	01:41:18.1	F	F
SARSAT-12	14.1095	856.25	01:42:03.5	F	F

of 406 MHz beacons, but since the geostationary satellites are by definition stationary with respect to the Earth, there is no Doppler shift of the received beacon carrier; consequently, there is no location determination. Another issue is that GEOs cannot cover polar regions since the antenna footprint is limited to latitudes of about 75–80° (Richharia 1999; Roddy 2006).

LEO satellites in polar orbits cover these potential distress regions, and Doppler shift processing is applied for location determination. Thus, the supplemental activity of both LEO and GEO satellites enables global coverage, and the use of both reduces the crucial time required to determine location, enabling faster rescue. Table 8.1 presents the orbital parameters and the status of SARSAT LEOSAR payload instruments (NOAA-SARSAT 2022) Global is related to global coverage, whereby F means fully operational.

For search and rescue missions, LEO satellites are equipped with a Search and Rescue Repeater (SARR) or Search and Rescue Processor (SARP), applied for two modes of operation (NOAA-SARSAT 2022; International COSPAS-SARSAT program 2013). The main functions of SARP are:

- Measure the signal's frequency.
- Time tag the frequency measurement.
- Extract the beacon ID.
- Convert the 406 MHz beacons uplink into a downlink data stream.

There are two modes of operation, local and global mode:

1) Repeater mode (local). SARR immediately retransmits received beacon signals to any LUT in the satellite's footprint. This mode is possible when the spacecraft is visible to both the beacon and ground station simultaneously, as in Figure 8.4 (NOAA-SARSAT 2022).
2) Store and forward mode (global). Applied when the spacecraft does not see the beacon and ground station simultaneously. This mode is presented in Figure 8.5 (NOAA-SARSAT 2022). The on-board SARP receives and records search and rescue beacon transmissions and repeatedly retransmits them to LUTs when they are visible as the satellite orbits the Earth. SARP processor stores and forwards the information from a continuous memory dump, thus providing complete worldwide coverage. Adding a processor, compared with poor repeater, reduces the mean waiting time for distress location determination (NOAA-SARSAT 2022).

In both modes, a satellite communicates on a downlink carrier frequency of 1544.5 MHz to one of several LUTs. The LUTs are fully automated and completely unmanned at all times (Landis and Mulldolland 1993; Golshan et al. 1996; Cakaj and Malaric 2007a).

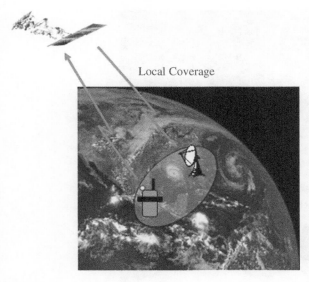

Figure 8.4 Local mode.

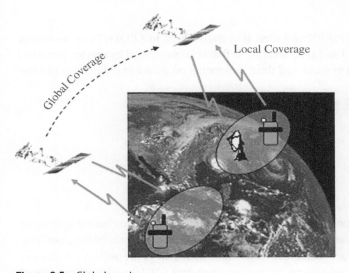

Figure 8.5 Global mode.

8.2.2 SARSAT Ground Segment

Receive-only ground stations specifically designed to track the search and rescue satellites as they pass across the sky are called LUTs, as shown in Figure 8.6 (NOAA-SARSAT 2022).

There are two-types of LUTs: low Earth orbiting local user terminals (LEOLUT) and geostationary local user terminals (GEOLUT). These LUTs and appropriate MCC to whom these LUTs are interconnected create the US SARSAT ground segment. The distress signal is received by satellite uplink (emergency locator transmitter (ELT), personal locator beacon (PLB), and emergency

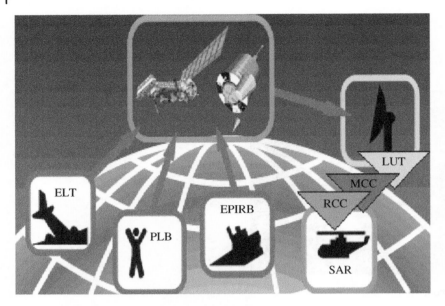

Figure 8.6 Local user terminal (LUT) and different beacons.

position indicating radio beacon (EPIRB)) and then it is transmitted to LEOLUTs by downlink (indicated by arrows in Figure 8.6). The LUT's location is fixed (implemented by service provider) and the beacon's location is random in space and time, depending on where and when the distress happens.

The main functions of a LEOLUT are:

- Track the SARSAT satellites.
- Recover beacon signals.
- Perform error checking.
- Perform Doppler processing.
- Send alert to MCC.

The LEOLUT system usually includes a satellite receive antenna, a digital processing system, an operator display mode and the software which implements all of the control, monitoring, and processing functions. Since LEOLUTs track satellites in low orbits, which move too quickly relative to a fixed point on Earth, the antenna includes an antenna control unit (ACU) and tracking mount mechanism with azimuth range of 360° and elevation up to 90°. LEOLUT satellite dishes are preprogrammed to move to pick up a LEOSAR satellite signal as the satellite passes overhead. The appropriate antenna software controls the pointing of the antenna.

When a satellite receives a beacon signal from a distress location, the SARP on board of LEO satellite performs Doppler processing and generates a 2.4 kb/s processed data stream (*pds*) to LEOUT, as in Figure 8.7 (COSPAS-SARSAT local user terminals 2012). LEOLUT software accepts the data stream, decodes, and extracts beacon data messages. From satellite pass collected by LEOLUT, software selects data from each detected beacon and validates time, frequency, and message content.

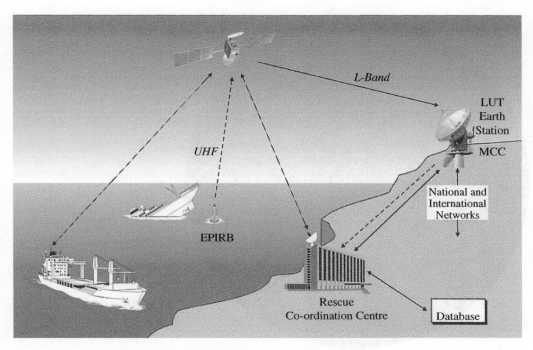

Figure 8.7 COSPAS-SARSAT local user terminals.

Data from each pass are then passed to the solution processing software. The solution processing software determines an optimum location based on Doppler frequency curve. Once a signal is processed at the LUT, the data stream that provides status (data about distress location) is transmitted through a fully automatic communication link to the MCC (NOAA-SARSAT 2022; International COSPAS-SARSAT program 2013). A MCC serves as the hub to collect, store, and sort alert data from LUTs and other MCCs. The main function of an MCC is to distribute alert data to RCCs and other MCC. TheUSMCC in Suitland, MD, serves as the focal point of the US SARSAT program.

NOAA operates 11 LEOLUTs in six locations, as presented in Table 8.2 (NOAA-SARSAT 2022) and Figure 8.8. These multiple LEOLUTs provide total system redundancy and allow for a

Table 8.2 LEOLUTs coordinates.

LEOLUTs locations	Latitude	Longitude
Maryland (MDLUT)	38.852	−76.937
Florida (FLLUT)	25.61	−80.38
California (CALUT)	34.66	−120.55
Alaska (ALLUT)	64.97	−147.51
Hawaii (HILUT)	21.52	−151.99
Guam (GULUT)	13.34	144.56

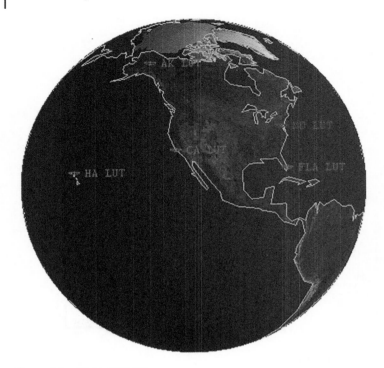

Figure 8.8 NOAA LEOLUTs.

maximization of satellite tracking. There are two LEOLUTs in each of the following locations. Two independent, yet functionally and physically identical, systems manufactured by EMS Technologies, a Canadian company, are implemented. Since each LUT operates independently, they are denoted as 1 and 2. The LEOLUT in Maryland is used for test purposes.

- Miami, Florida (FL1&FL2)
- Vandenberg, California (CA1&CA2)
- Fairbanks, Alaska (AL1&AL2)
- Wahiawa, Hawaii (HI1&HI2)
- Andersen, Guam (GU1&GU2)
- Suitland, Maryland (LEOLUT support equipment)

8.2.3 Beacons

Emergency distress beacons are essentially specialized radio transmitters for search and rescue purposes. Satellite-aided search and rescue (SAR) satellites can "hear" even faint distress signals from beacons. Beacons can be activated manually or automatically. Since February 2009, all these rescue beacons have transmitted on 406 MHz. Emergency beacons are classified as (NOAA-SARSAT 2022; COSPAS-SARSAT. Int 2020):

1) EPIRBs for maritime applications
2) ELTsfor aviation applications
3) PLBs for individuals

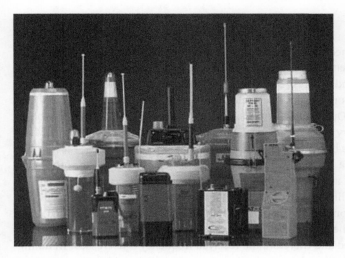

Figure 8.9 Beacons.

Table 8.3 406 MHz beacons characteristics.

Output power	5 W
Transmission	Burst (500 ms on every 50 s)
Modulation	Phase
Frequency stability	High

Different types of beacons are presented in Figure 8.9.

Some characteristics of 406 MHz beacons are shown in Table 8.3 (Specification for COSPAS – SARSAT406MHz, 2008).

The transmitted power output must be within limits of 5W ± 2dB. Omnidirectional antenna defined for all azimuth angles and for elevation above 5° is used. Frequency spectrum lies in the band 406.0–406.1 MHz. SARSAT 406 MHz receivers are designed to detect beacon signals operating within the full designated bandwidth for low power satellite emergency position – indicating radio beacons of 406–406.1 MHz. An additional bandwidth capability is also needed to accommodate possible Doppler shift at both boundaries of the operating band.

Each receiver is equipped with filters capable of attenuating signal outside this band to avoid overloading the receiver with "out of band" signals. Because frequency stability is so important for determining location, the carrier frequency should not vary more than ±5 KHz from 406.025 MHz in five years (Specification for COSPAS – SARSAT406MHz 2008).

The 406 MHz carrier is modulated with information such as beacon identification, synchronization frame, and the nature of emergency. Looking at the time slot of 500 ms (Table 8.3), only 160 ms are dedicated for poor carrier and the rest is for modulated data (Specification for COSPAS – SARSAT406MHz 2008). Beacon identification transmission is mandatory for accessing a user registration database. This database can supply the beacon type, its country of origin, and the registration number of the maritime vessel, aircraft, or individual. US law requires that beacons be registered; there are approximately 270 000 registered beacons in use in the United States.

8.3 Doppler Shift

The communication link is established when the satellite flies over the LUT's visibility. This "fly-over" is called a *satellite pass*. LEO satellites move at around 7.5 km/s ensuring sufficient velocity relative to a fixed point on the ground (distress location) to generate a perceptible Doppler shift in the frequency of the emergency beacon signal, thereby yielding reasonably accurate position of distress location (NASA Final Report, Operational SAR Doppler Processing System 1981). The basic system concept requires users to carry distress radio beacons, which transmit a carrier signal when activated. The transmitted signal by a beacon is picked up by LEO satellite, and because the satellite is moving relative to radio beacon, a Doppler shift in frequency is observed, and the shifted frequency and time tag is registered at a satellite processor (SARP) (Taylor and Vigneault 1992).

What is the Doppler shift? If the line-of-sight distance between the transmitter and satellite is shortened as a result of relative motion, the wavelength of the emitted signal is also shortened, consequently frequency is increased. If the line-of-sight distance is lengthened as a result of the relative motion, the wavelength is lengthened and therefore the received frequency is decreased. Denoting the constant emitted frequency by f_0, the relative velocity between satellite and beacon measured along the line of sight as v, and the velocity of light as c, then to a close approximation the received frequency at the satellite, is given by Richharia (1999) and Roddy (2006):

$$f = \left(1 + \frac{v}{c}\right)f_0 \tag{8.1}$$

The relative velocity v is positive when the line-of-sight distance is decreasing (satellite and beacon moving closer together) and negative when it is increasing (satellite and beacon moving apart). The relative velocity v is a function of the satellite motion and of the Earth's rotation, and can be approximated as (Richharia 1999; Roddy 2006):

$$v = \frac{R(T_X) - R(T_{X-1})}{T_X - T_{X-1}} \tag{8.2}$$

where $R(T_X)$ and $R(T_{X-1})$ are satellite ranges at times T_X and T_{X-1} respectively, for $(T_X - T_{X-1})$ arbitrarily small. The frequency difference resulting from relative motion is (Richharia 1999; Roddy 2006):

$$\Delta f = f - f_0 = \frac{v}{c}f_0 \tag{8.3}$$

when v is zero, the received frequency is the same as the transmitted frequency. When beacon and satellite are approaching each other, v is positive, which results in a positive value of Δf. When the beacon and the satellite are receding, v it is negative, resulting in a negative value of Δf. The time at which $\Delta f = 0$ is known as the *time of closest approach*.

In all cases, the received frequency goes from being higher to being lower than the transmitted value as the satellite approaches and then recedes from the beacon. The longest record and the greatest change in frequency are obtained if the satellite passes perpendicularly over the site, because the satellite is visible for the longest period during this pass. Considering the orbital parameters of the satellite, the beacon frequency, and the Doppler shift for any pass, the distance of the beacon relative to the projection of the orbit on the Earth can be determined. The determination location is based on time of closest approach and on slope of Doppler curve. This is presented in Figure 8.10. The beacon of 406 MHz, when has view of site with the satellite, hits the satellite with bursts of 406 MHz for 500 ms, periodically every 50 s. The satellite processor records the frequency and time tag. This beacon's hit with recorded frequency and time tag represents a Doppler Event

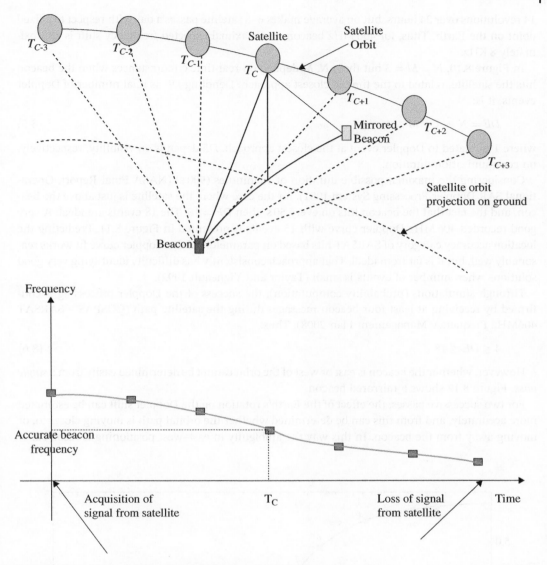

Figure 8.10 Doppler events and respective frequency shifts.

(*DE*). The number of Doppler events depends on visibility time between a beacon and satellite. Figure 8.10 illustrates the case with seven Doppler events. In Figure 8.10, T_C represents the time of closest approach. T_{C-N} denotes times of Doppler events before the closest approach and T_{C+M} times of Doppler events after the closest approach. At the bottom of Figure 8.10, frequency shifts of respective events are presented.

CIR Report 214 gives the following approximate relationship for estimating the maximum Doppler shift (Richharia 1999; Roddy 2006)

$$\Delta f_{dm} \approx \pm 3 \cdot 10^{-6} f_0 s \tag{8.4}$$

where f_0 is operating frequency and s is the number of revolutions over 24 hours of the satellite with respect to a fixed point on the Earth. A LEO satellite with an orbital period of 100 minutes makes

14 revolutions over 24 hours, but on average makes 6–8 satellite passes a day with respect to a fixed point on the Earth. Thus, for 406 MHz beacon, the maximal expected frequency shift is approximately 8 KHz.

In Figure 8.10, $N = M = 3$ but these N, M depend on real-time circumstances when the beacon hits the satellite, related to the time of closest approach. Denoting DE as total number of Doppler events, it is:

$$DE = N + 1 + M \tag{8.5}$$

where 1 is related to Doppler event at the closest approach. DE depends on visibility, respectively, on communication duration.

Considering the maximal possible duration of 15 minutes (900 s) (NASA Final Report, Operational SAR Doppler Processing System 1981), for the case where the satellite is just above the beacon, and the fact that the beacon hits on every 50 s, then $DE = 18$. The 18 events are ideal. A very good recorded 406 MHz Doppler curve with 15 events is presented in Figure 8.11. Predicting the location accuracy category of SARSAT hits based on parameters of the Doppler curve fit works reasonably well, but it is far from ideal. That approach consistently has difficulty identifying very good solutions when number of events is small (Taylor and Vigneault 1992).

Through simulations (probability computation), the success of the Doppler processing is confirmed by receiving at least four beacon messages during the satellite path (COSPAS – SARSAT 406MHz Frequency Management Plan 2008). Thus:

$$4 \leq DE \leq 18 \tag{8.6}$$

However, whether the beacon is east or west of the orbit cannot be determined easily from a *single pass*. Figure 8.12 shows a mirrored beacon.

For two successive passes, the effect of the Earth's rotation on the Doppler shift can be estimated more accurately, and from this can be determined whether the orbital path is moving closer to, or moving away from the beacon. In this way the ambiguity in east–west positioning is resolved.

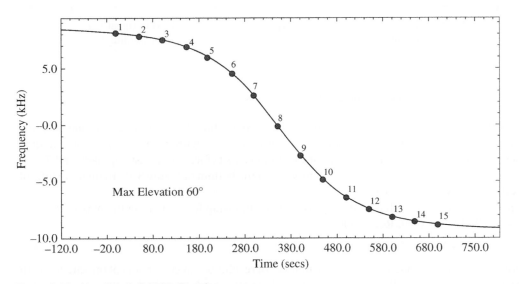

Figure 8.11 Very good 406 MHz Doppler curve.

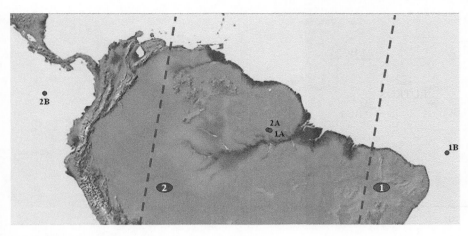

Figure 8.12 Ambiguity solution.

Figure 8.12 describes the solution for the ambiguity. During the first pass, (1) processing identify locations 1A and 1B, and during the second pass processing are identified locations 2A and 2B. It is very clear from Figure 8.12 that the distress location was in vicinity of 1A and 2A (USTTI Course M6-102 2006).

8.4 Local User Terminal (LUT) Simulation for LEO Satellites

LUTs for LEO satellites, known as LEOLUTs, are part of an international communication network dedicated for search and rescues service worldwide (given in Figure 8.1).

Idea: Since the LUTs number is increasing making the appropriate network safer and better covered, the further idea is the simulation for a new LUT implementation at the hypothetical site. LUT is assumed to be implemented in Republic of Kosovo and considered for further simulation is defined as LUTKOS (local user terminal in Kosovo) (Cakaj et al. 2010). The idea is, the communication reliability between satellites and LUTs will be confirmed.

Method: Simulation is applied. For SARSAT system, the uplink transmitter is distress beacon, and the LUT is downlink receiver. Thus, the proper operation for search and rescue services should be analyzed for uplink as seen from random beacons appearance (distress events) and for downlink as seen from the satellite to the fixed LUT on ground.

Four hypothetical beacons for analyses on uplink, and a fixed ground station LUTKOS for downlink are considered. Republic of Kosovo lies on these approximate coordinates: Latitude of $42°$ and longitude of $21°$. Under simulation assumption is considered the area with line of sight of (1100–1500) km from the center, respectively from LUTKOS. The area to be considered for simulation is presented in Figure 8.13 and data about LUTKOS and hypothetical beacons in Table 8.4 (Cakaj et al. 2010). This range provides coverage on the Earth's surface of at least 3 million square kilometers given in Figure 8.14.

The goal of simulation is to confirm reliable data communication between assumed LUTKOS ground station and SARSAT satellites, and to confirm the visibility of hypothetical beacons with SARSAT satellites in terms of Doppler events compliance (Eq. (8.6)). These two facts will prove the proper operation of LUTKOS dedicated for search and rescue services. Single satellite is

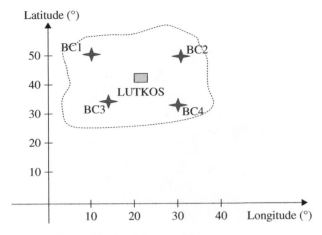

Figure 8.13 LUTKOS and beacons idea.

Table 8.4 Coordinates of LUTKOS and beacons.

Location	Latitude	Longitude
LUTKOS	42.30	21
BC1	50	10
BC2	50	30
BC3	34	30
BC4	34	15

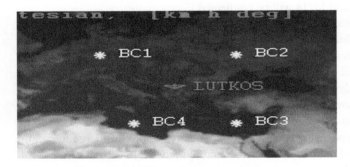

Figure 8.14 LUTKOS and beacons simulated.

Figure 8.15 Satellite, LUTKOS, and beacons.

considered for simulation. The presence of more satellites improves results. For simulation purposes the appropriate NOAA software is used.

Practically, this means that at least high Alps Mountains in Austria and very large area of Mediterranean Sea to be covered with search and rescue services.

NOAA environmental satellite with SARP for further simulation is added as presented in
Figure 8.15. Satellite's orbit is considered as circular with no eccentricity, the orbital altitude of
860 km, orbital time of 102 minutes and Inclination of 98.7°. The satellite's antenna conic angle
is 60° (Cakaj et al. 2010; Cakaj 2010; Cakaj 2014).

A single satellite is considered for simulation. The presence of more satellites improves results.
Here we consider different relative positions among satellite, LUTKOS, and beacons to examine
operation and service performance. Three typical cases in Figures 8.16–8.18 are further presented:

1) During the satellite pass (presented in Figure 8.16) beacons BC1 and BC4 are within a satellite
   footprint. Two other beacons (BC2 and BC3) are out of footprint. LUTKOS is within a footprint.
   In this case, the satellite can receive signals from beacons BC1 and BC4. The satellite can transmit the signal to LUTKOS, and LUTKOS further to respective MCC. Search and rescue action
   can then be taken. Eventually, distress signals from BC2 and BC3 should wait for the next pass or
   another satellite.
2) During the satellite pass (presented in Figure 8.17), all beacons and LUTKOS are within a footprint. From the search and rescue view, this is the most optimistic case for search and rescue
   services. Each beacon can communicate with the satellite and the satellite can download data
   to the LUTKOS.

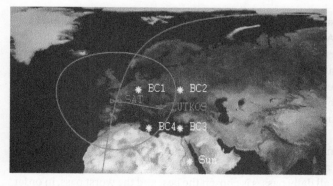

Figure 8.16 The first case.

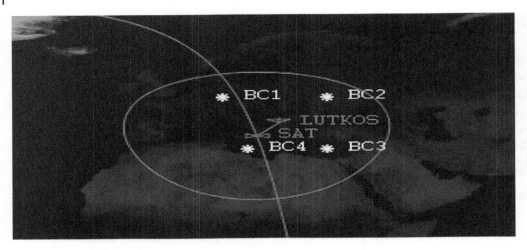

Figure 8.17 The second case.

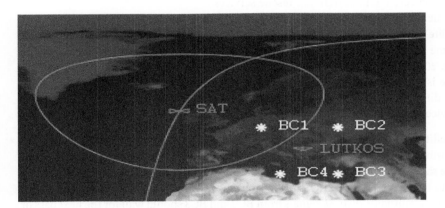

Figure 8.18 The third case.

3) During the satellite pass (presented in Figure 8.18), only Beacon BC1 is within a satellite footprint. LUTKOS and other beacons are out. This is the pessimistic case, since the distress beacon communicates with the satellite but the satellite cannot communicate with LUTKOS, so the satellite should look for another LUT to download data. This is a delay.

These three examples confirmed insufficiency of a single satellite; thus six NOAA LEO environmental satellites are equipped with SARP.

Results: Analyses are related to a period of one month from October 1 to October 30, 2009. Maximal Elevation angle and Communication Duration is conducted. Appropriate software provides these data in different formats, tabulated, or graphically, which can further be processed with Excel. Maximal elevation variation during the period (October, 1 to October 30) between LUTKOS and SAT is presented in Figure 8.19. Higher maximal elevation provides longer communication between LUTKOS and satellite.

Figure 8.20 shows results of communication duration between the LUTKOS and satellite. For the whole period under conduction, data are gathered for further processing as Excel files. For Excel presentation of the whole period, from all daily passes is chosen the best and the worst pass, in order

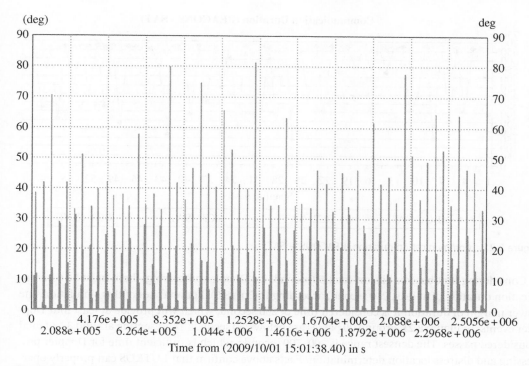

Figure 8.19 Maximal elevation.

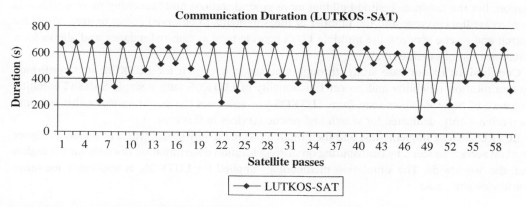

Figure 8.20 Communication duration (LUTKOS − SAT).

to create real opinion about the communication duration time range. Figure 8.20 shows that only one pass is under 200 seconds, few around 300 seconds, but the highest density is concentrated in the range of 400–700 seconds, or between 6 and 12 minutes. This communication duration well satisfies data download from the satellite to LUTKOS.

The same approach is conducted for all beacons, so the best and the worst pass is considered, and then presented in Figure 8.21. All beacons within a period of one month are treated around 700 passes (Cakaj 2010; Cakaj 2014). Successful Doppler processing is defined by reception of at least four beacon messages (conform Eq. (8.6)) during the satellite path.

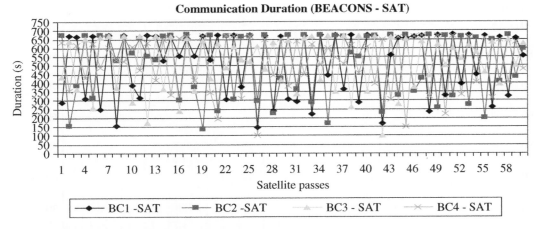

Figure 8.21 Communication duration (BEACONS – SAT).

Considering that beacon hits on each 50 second, for Doppler processing, there must be communication duration of at least 200 second, according to Eq. (8.6). For analyses of this simulation, the lower margin of duration it is considered time of 250 second, which will provide at least four Doppler events. In Figure 8.21, around 20 passes are below 250 seconds, representing around 3% of total considered passes. The densest range is 300–700 seconds, which is sufficient time for Doppler processing and distress location determination. Facts above confirm that LUTKOS can properly operate and be a part of LUTs network (Cakaj 2010; Cakaj 2014).

In principle, any satellite mission can be accomplished by a single satellite and single ground station, but the rationale behind building more ground stations and launching more satellites is to increase the coverage and the number of measurements to observed object or area. Thus, for search and rescue services, the multiple LUTs provide total system redundancy and allows for a maximization of satellite tracking.

Conclusion: In our case study, through simulation approach for LUTKOS, it is confirmed communication reliability and proper functionality of LUTKOS with a single SARSAT satellite. Presence of more satellites seen from LUTKOS just enhances the performance of this terminal as receive – only, dedicated for search and rescue services in this area.

LUTKOS can be implemented and then interconnected to any of the MCCs in region as higher level on service concept, by contributing on distress location determination not only for the region but also worldwide. The simulation methodology applied for LUTKOS, is applicable for other worldwide sites, also.

8.5 Missed Passes for SARSAT System

In principle, any satellite mission can be accomplished by a single satellite and single ground station, but the rationale behind building more ground stations and launching more satellites is to increase the coverage and the number of measurements to observed object or area. US SARSAT is a data communication system dedicated for search and rescue purposes oriented on determination of distress locations worldwide, where one location is unpredictable in ground area and in time. Thus, for search and rescue services, the multiple LUTs provide higher system redundancy and allows for a maximization of satellite tracking.

Even under too high coverage, because of natural barriers, could happen that the communication between beacons at distress location cannot to be locked in uplink with the satellite, or if it is even established with too short communication, will not be provided the sufficient Doppler events. Also, on downlink because of too-low elevation, if the LUT cannot be locked in with the satellite, the data cannot be downloaded from the satellite to the LUT. This means that if there is a delay in determining distress location determination due to inefficient systems, human lives are at risk!

In order to avoid the problem of natural barriers, designers predetermine the lowest elevation of the horizon plane (Cakaj 2013). Considering a safe margin, this elevation ranges from 5° to 30° (Essex et al. 2004). The horizon plane with a predetermined minimal elevation is considered the *designed horizon plane*. For LUTs of SARSAT, the designed elevation is 5° (NOAA–SARSAT 2022). Thus, from the communication point of view, beacons are dynamic points on Earth with unknown coordinates (unpredictable when and where happens). LUTs are fixed with exactly known position, coordinates. Both, LUTs and beacons, from analytical point of view are access points (ground stations) in relation to the satellite, beacons communicating on uplink and LUTs on downlink. Because of beacons unpredictability, further is discussed the satellite's access in terms of providing the sufficient Doppler events, for distress location determination.

Beacons can be considered as uplink ground stations, whose position is not initially known. This position represents the distress location. Beacons can communicate with the satellite only when it is within the satellite's coverage zone. For LEO satellites, the coverage zone is in a range of 5000–6000 km in diameter, depending on the altitude and the elevation a satellite is flying. The beacon can be at any point on the Earth's surface simply characterized with its longitude and latitude. Further, three beacons are considered and represented with their respective horizon planes. A single satellite is considered. The case under this assumption is presented in Figure 8.22. During the satellite pass, the Beacon 1 sees the satellite under low maximal elevation (*MaxEl* of 12°) and thus has too short communication time to provide enough Doppler points for distress location determination. Satellite passes by the edge of horizon plane of the Beacon 2, having no communication with it. Beacon 2 should be waiting for the next pass or for another satellite. There is no reception from Beacon 2. Beacon 3 sees the satellite almost over its head, under high maximal elevation (*Max-El* of 85°) having enough long communication with the satellite for distress location determination.

There is a clear relationship between communication duration and maximal elevation. From commercial LEO SARSAT software, the relationship between maximal elevation and communication duration for satellite daily passes is presented in Figure 8.23, where higher peak represent longer communication, and longer communication provides more Doppler events, and thus better accuracy for determining location. The satellite pass, which does not provide enough events for Doppler solution, is considered as a *missed pass* (in Figure 8.22, satellite pas over Beacon 1).

Idea: The intention was to analyze missed passes caused by hardware/software malfunctioning at LUTs and missed passes under too-low elevation, which do not provide Doppler solutions.

Method: Records are analyzed and processed related to each site/month and each site/satellite, for eight months. Summarized results are presented in Tables 8.5 and 8.6 (Cakaj 2010; Cakaj 2014).

Results: Both tables show that for the considered period of eight months the US SARSAT ground segment communicating with six satellites had 98 missed passes. This means, on average, 1.225 missed passes per site/month with all satellites, or in average 0.24 missed per site/month/satellite. The total number of passes during this period of eight months was 57 560, and the ratio of missed passes over total number of passes for the whole ground segment is 0.17%, or in average 0.021% per month. Thus, monthly ground segment performance indicator is 99.979% (Cakaj 2010, Cakaj 2014). This confirms high performance indicator of SARSAT system from the missed passes perspective.

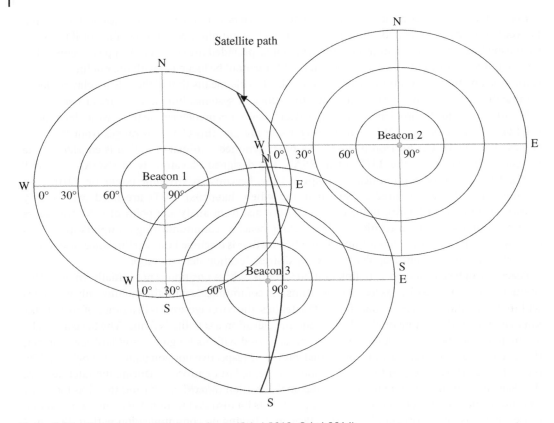

Figure 8.22 Three beacons' horizon planes (Cakaj 2010; Cakaj 2014).

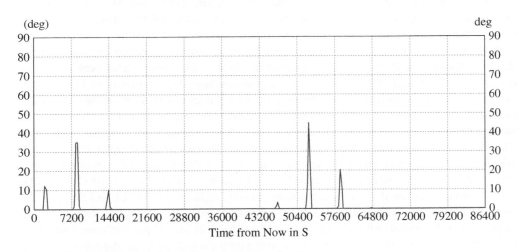

Figure 8.23 Maximal elevation and communication duration.

Table 8.5 Missed passes as site/month.

	Feb.	March	Apr.	May	June	July	Aug.	Sep.
CA1	0	1	0	1	0	2	1	1
CA2	0	1	1	2	2	1	0	1
HI1	0	1	0	0	1	0	1	3
HI2	0	0	3	0	0	0	2	1
FL1	1	1	1	0	2	1	2	5
FL2	0	0	1	0	0	0	1	2
GU1	0	1	0	0	3	4	1	0
GU2	0	0	2	3	4	0	2	3
AL1	3	0	0	0	6	1	3	0
AL2	0	2	0	3	3	1	5	4

Table 8.6 Missed passes as site/satellite.

	S7	S8	S9	S10	S11	S12
CA1	1	0	3	1	0	1
CA2	1	0	3	2	1	1
HI1	1	0	0	3	1	1
HI2	1	0	2	2	1	0
FL1	0	2	3	3	2	3
FL2	0	1	1	1	2	1
GU1	3	0	1	3	0	2
GU2	2	0	2	6	2	2
AL1	1	4	3	2	2	1
AL2	1	2	4	3	6	1

This performance guarantees on time alert to rescue services and aides' lifesaving rescue efforts. The implementation of the SARP global mode enables PDS data capturing at multiple sites covering also missed passes. The data capture redundancy ensures uninterrupted service of a critical lifesaving system.

Conclusion: Through LUTKOS simulation, it is confirmed communication reliability and proper functionality of LUTKOS with a single SARSAT satellite. More SARSAT satellites seen from LUTKOS enhance the performance for search and rescue services.

DASS (Distress Alert Satellite Mission) is a new approach intended to enhance the international SAR system by installing 406 MHz SAR instruments on the medium Earth orbit (MEO) navigational satellites (GPS [US], Galileo [EU], and Glonass [Russian Federation]) and by introducing new processing algorithms.

8.6 LEOSAR Versus MEOSAR

LEOSAR detects beacons only when that beacon is physically under the LEO satellite's footprint, which is relatively small, compared to the Earth's surface. Further processing is based on Doppler calculations. If the number of Doppler events, or bursts, is less than 4 (the case of too-short visibility above the beacon's location), the calculation is difficult and will lead to larger error margin. This increases inaccuracy on distress location determination. If the beacon is not within the LEOSAR footprint, the 406 MHz burst cannot be detected or located until the appropriate satellite footprint covers the beacon's area, consequently resulting in a delay on location determination. Thus, LEO-SAR has limited instantaneous coverage. Another problem with LEOSAR is the signal blockage caused by terrain obstruction and masking under low elevation.

To improve the performance, this system is thought to be migrating toward MEOSAR (medium Earth orbit search and rescue), restructuring the capability to MEO satellites. By 2005, NASA had the MEOSAR Implementation Plan, then "Proof of Concept" was written by 2010 and moving forward by installation of devices, so by 2016, it had six MEOLUT antennas, located in Hawaii and Florida. Recently, LEOSAR and MEOSAR are cooperating since MEOSAR it is not yet complete (MEOSAR overview 2016). The MEOSAR operation concept remains similar as LEOSAR, but MEOSAR includes SAR payloads on three global navigation satellite systems (Galileo, GLONASS, and GPS) (International COSPAS-SARSAT program 2013; NOAA-SARSAT 2022) as given in Figure 8.24.

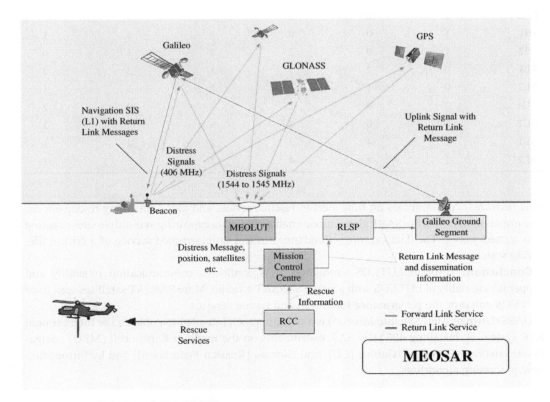

Figure 8.24 COSPAS SARSAT (MEOSAR) concept.

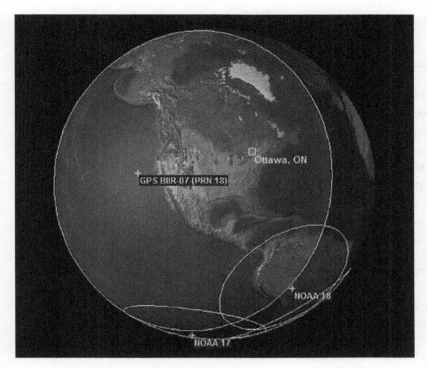

Figure 8.25 MEO and LEO coverage comparison.

A full MEOSAR constellation will contain numerous satellites. Each MEO satellite has a much larger footprint than a LEOSAR satellite. The coverage area increases with satellite's altitude, and MEO satellites orbit on 19 000–23 000 km versus LEO satellites at range of 600–1000 km. With enough MEOLUTs and MEOSAR satellites, coverage will be possible at any point on Earth. This will enable instantaneous worldwide coverage. Figure 8.25 shows the coverage of a typical GPS satellite and two NOAA LEO satellites (Comparison of LEO and MEO footprints, CRC Canada 2016). This difference in coverage enables MEOSAR satellites to detect a much larger area. This, coupled with the fact that there are already 20 GPS satellites and 10 Galileo satellites equipped with MEOSAR payloads allows the beacons to be located more quickly from MEOSAR than from the LEOSAR constellation.

Just a single beacon burst will be sufficient for MEOSAR detection. Since multiple links are used, no Doppler curve is needed. The LEOSAR concept requires at least four bursts to build a Doppler curve for location determination. As the MEOSAR system will be able to provide a single-burst location, the frequency stability is less important than for LEOSAR.

Idea: The performance analysis of the currently operational LEOSAR system and the future planned MEOSAR system is viewed in terms of visibility and availability. Then, specifically in both cases, the coverage is issued under the circumstances considered for the simulation (Kamo et al. 2018).

Method: For the comparison purposes, live satellites tracking information of the LEOSAR and GPS constellations (US MEOSAR) for over a period of 24 hours is used. The tracking is made possible using satflare.com and the ephemeris data of the Almanac YUMA file on July 2, 2017. The LEOSAR system satellite constellation of six LEO satellites and the GPS satellites are observed from the hypothetical geographical location (coordinates of N 40.90028°, E 19.91667° and altitude of 100 m

Table 8.7 Satellite passes of the LEOSAR system for the chosen location, during 24 hours.

Satellite	Rise	Culminate	Set	Best time	Sat. Elev. at best	Duration
NOAA 15	–	–	–		–	–
NOAA 16	22:42:33	22:50:24	22:58:15	22:49:45	58^0	16 min
NOAA 18	20:51:09	20:56:48	21:02:26	20:54:49	7.7^0	11 min
METOP A	20:29:26	20:36:27	20:43:28	20:35:59	25.5^0	14 min
	22:08:28	22:16:19	22:23:49	22:15:19	37.7^0	15 min
	23:54:58	23:57:40	00:00:21	23:55:31	0.5^0	6 min
NOAA 19	02:42:37	02:48:33	02:50:29	02:50:19	9.2^0	8 min
	04:22:04	04:29:57	04:37:50	04:30:12	83.5^0	15 min
METOP B	21:22:33	21:30:14	21:37:55	21:29:51	70^0	15 min
	23:04:34	23:10:50	23:17:07	23:09:29	12.6^0	13 min
TOTAL						113 min

above the sea level, Tirana city, Albania) from where eventually the emergency beacon signal will be generated. Further processing of the Almanac YUMA file is run using Trimble's Planning Software (Kamo et al. 2018).

Results: Table 8.7 provides the live satellite tracking of the LEOSAR system from the hypothetical location, including the visibility duration.

Results show that the visibility of the beacon at the hypothetical location, from the above listed LEOSAR satellites, is relatively low. The total communication duration with the LEOSAR system is no more than 113 minutes per 24 hours, giving an availability of about 7.8% of the considered time interval and consequently causing delays on location determination, respectively, on time to access at location of the distress event (Kamo et al. 2018).

For Doppler processing, at least four events or bursts are required. If this criterion is violated, in case of too short visibility with the beacon or the blockage from the terrain obstruction causing the masking under the low elevation, then the inaccuracy in location determination and the error margin increases. Under the above observation, the most critical case is evidenced for the satellite passes of METOP A, having communication duration of 6 minutes (Table 8.7). This short pass, even under the case that it could provide more than four Doppler events, the low number of events, seriously impacts the accuracy on location determination.

Further, using the live tracking, the satellite's visibility is recorded from the hypothetical beacon location for both LEOSAR and MEOSAR (GPS), as given in Figure 8.26, which shows that the communication duration of only one GPS (MEO) satellite varies from one to a few hours, compared to a total of 113 minutes of communication duration of all the LEOSAR system. The considered interval between 20:00 and 24:00 was preferred in order to emphasize the difference in real time of both systems when the LEOSAR system is at most available (Kamo et al. 2018).

Figure 8.27 further emphasizes the differences in availability. The basic principle of satellite-based navigation requires having at least four GPS satellites in line-of-sight to the beacon (MEOSAR case). The results in Figure 8.27 show that the beacon is continuously seen from at least five GPS satellites thus meeting the criteria. From Figures 8.26 and 8.27 the advantage in visibility and availability of MEOSAR compared with LEOSAR, it is obvious (Kamo et al. 2018).

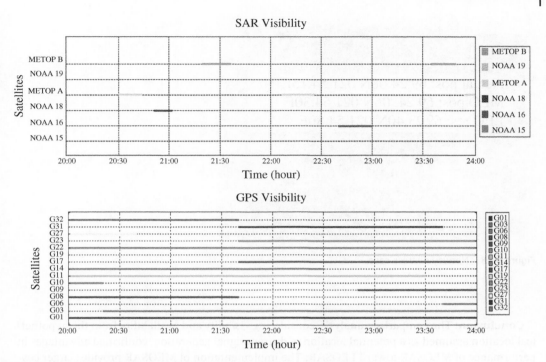

Figure 8.26 MEOSAR vs. LEOSAR system visibility during 20:00 to 24:00.

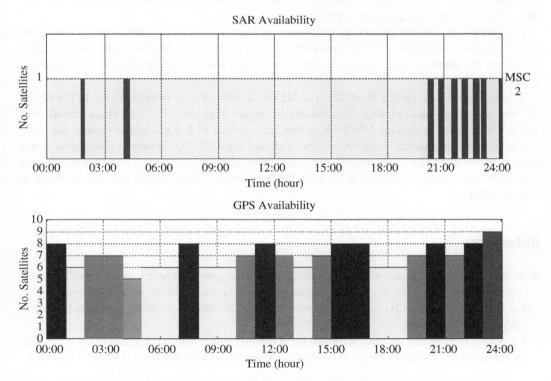

Figure 8.27 Number of MEOSAR vs. LEOSAR system satellites during 24 hours seen from beacon.

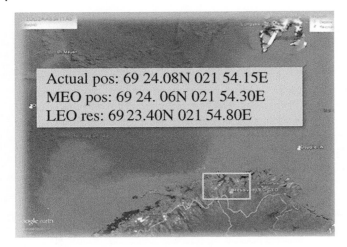

Actual pos: 69 24.08N 021 54.15E
MEO pos: 69 24. 06N 021 54.30E
LEO res: 69 23.40N 021 54.80E

Figure 8.28 Norway distress case.

Conclusion: The comparison analysis, considering visibility and availability from the hypothetical location assumed as a potential location of distress signal generation, confirmed advantages in performance of MEOSAR toward LEOSAR. The implementation of MEOSAR provides larger coverage and instantaneous alert, compared to LEOSAR. Thus, the MEOSAR location estimation is much faster than with LEOSAR. A quicker distress alert directly contributes to more effective rescue services where timing is usually critical.

This section, to confirm the above conclusion, is closed with the case happened in Norway on 15 August 2016. LEOSAR resolved the location by 1.5 km far from the crash site, but MEOSAR confirmed the position only 100 m far from the crash site (Generic MEOSAR presentation 2016) given in Figure 8.28.

Data in Figure 8.28, clearly show that the MEOSAR calculated position is closer to the actual distress position, compared with the calculated position from the LEOSAR. These results may not be typical for the current MEOSAR system, due to lack of full space and ground segment, but when fully operational capability will be achieved, the MEOSAR system is going to be much more accurate as the LEOSAR system, while providing global coverage and real time search and rescue services. Thus, the migration from LEOSAR toward MEOSAR search and rescue system is fully justified.

References

Bulloch, C. (1987). Search and rescue by satellite – slow steps toward an operational system. *Interavia*, (ISSN 0020-5168) 42: 275–277. https://ui.adsabs.harvard.edu/abs/1987Inter..42..275B/abstract.

Cakaj, S. (2010). *Local User Terminals for Search and Rescue Satellites Book*, 84. Saarbrucken, Germany: VDM Publishing House.

Cakaj, S. (2013). Elevation variation with low Earth orbiting search and rescue satellites for the station implemented in Kosovo. *Universal Journal of Communications and Network* 1 (1): 32–37. CA, USA.

Cakaj, S. (2014). Communication duration and missed passes among terminal and satellites for search and rescue services. *British Journal of Mathematics & Computer Science* 4 (12): 1771–1785.

Cakaj, S. and Malaric, K. (2007a). Rigorous analysis on performance of LEO satellite ground station in urban environment. *International Journal of Satellite Communications and Networking* 25 (6): 619–643. UK, November/December 2007.

Cakaj, S., Fitzmaurice, M., Reich, J., and Foster, E. (2010). *Simulation of Local User Terminal Implementation for Low Earth Orbiting (LEO) Search and Rescrue Satellites*, 140–145. Athens, Greece: The Second International Conferences on Advances in Satellite and Space Communications, SPACOMM 2010, IARIA.

Comparison of LEO and MEO footprints, CRC Canada (2016), Next Generation Satellite Payloads in the MEO (Medium Earth Orbit) Space https://directory.eoportal.org/web/eoportal/satellite-missions/c-missions/cospas-sarsat.

COSPAS – SARSAT 406MHz Frequency Management Plan (2008) C/T T.012, Issue 1 – Revision 5, Probability of Successful Doppler Processing and LEOSAR System Capacity.

Cospas-Sarsat Local User Terminals, 2012; https://www.bing.com/images/search?view

Essex, E.A., Webb, A.P., Horvath, I. et al. (2004). *Monitoring the Ionosphere/Plasmasphere with Low Earth Orbit Satellites: The Australian Microsatellite FedSat*. Cooperative Research Center for Satellite Systems: Department of Physics, La Trobe University, Bundoora, Australia.

Generic MEOSAR presentation (2016) COSPAS-SARSAT secretariat, Montreal, Canada.

Golshan, N., Raferly, W., Ruggier, C. et al. (1996). Low Earth orbiter demonstation terminal. *TDA, Report* 42–125: 1–15.

International COSPAS-SARSAT program, 2013. https://www.icao.int/Meetings/GTM/Documents/COSPAS-SARSAT.pdf

Kamo, B., Jorgji, J., Cakaj, S. et al. (2018) The performance comparison for low and medium Earth orbiting satellite search and rescue services. *Advances in internet, Data & web Technologies, the 6th international conference on emerging internet, Data & web Technologies (EIDWT-2018)*, Albania, pp. 1–13.

Landis, S.J. and Mulldolland, E.J. (1993). Low cost satellite ground control facility design. *IEEE, Aerospace & Electronic Systems* 2 (6): 35–49.

Ludwig, D., Wallace, R., and Kaminsky, Y. (1985). *Proposed New Concept for an Advanced Search and Rescue Satellite System*. Stockholm, Sweden: IAF, 36th International Astronautically Congress.

MEOSAR Overview 2016, https://www.dco.uscg.mil/Portals/9/CG-5R/EmergencyBeacons/2016SarsatConf/Presentations/SAR_2016_Mar%201_%20MEOSAR%20Overview_LeBeau.pdf

NASA Final Report, Operational SAR Doppler Processing System (1981). Techno – Sciences Number 811130, NASA Goddard Space Flight Center Greenbelt, Maryland, USA.

NOAA-SARSAT, 2022; https://www.sarsat.noaa.gov

Richharia, M. (1999, 1999). *Satellite Communications Systems*. New York: McGraw Hill.

Roddy, D. (2006). *Satellite Communications*, 4e. New York: McGraw Hill.

Specification for COSPAS – SARSAT406MHz Distress Beacons, (2008), C/T T.001, Issue 3 – Revised.

SRC: COSPAS -SARSAT system, 2012; https://gmdsstesters.com/images/radio-survey/cospas-sarsat-coverage.jpg.

Taylor, W.I. and Vigneault, O.M. (1992). *A Neural Network Application to Search and Rescue Satellite Aided Tracking (SARSAT)*, 189–201. The Symposium/Workshop on Applications of Experts Systems in DND, Royal Military Coll. of Canada.

USTTI Course M6-102 (2006). *Telecommunications and Information Administration*. Washington, DC: Radio Frequency Spectrum Management Course.

9

Interference Aspects

9.1 General Interference Aspects

Generally, interference sources are statistically independent. Interference may be considered as a form of noise. Thus, the interference at each source may be added directly to give the total interference at the end user receiver. The effects of interference must be assessed in terms of: (i) the amount of frequency overlap between the interfering spectrum and the wanted channel passband of the end user receiver and (ii) the tolerable disturbing (interference) signal power level to the end user receiver. There are various sources of interference in a satellite communication system. These may be broadly classified as *intra-system* and *inter-system*.

Intra-interference can occur when the filters used for isolating adjacent channels do not have sufficient roll-off characteristic. This interference is called adjacent channel interference (ACI). Such interference can be minimized by using adequate guard bands between adjacent channels and well-designated filters. A *guard band* (GB) is the difference between the upper edge of the band and the last frequency within a band. However, the use of wide guard bands leads to inefficient use of the channel bandwidth and higher operating cost per carrier. Therefore, a technical and economic compromise should be made. Whatever the compromise is chosen, part of the power of a carrier adjacent to a given carrier will be captured by the receiver tuned to the frequency of the carrier considered. Quality is maintained if the captured power is under permitted limits. A link margin of (0.5–1) dB is adequate to compensate for intra-system interference (Richharia 1999).

Intra-system interference can be caused by coupling of orthogonally polarized signals in dual polarized system, also. A horizontally polarized signal can interfere with vertically polarized signal and vice versa. Rain and ice can cause interference at ground stations because of depolarization. This can be minimized by using well-designed ground station receivers and antennas (typical values of cross-polar discrimination in well-designed antennas are of the order of 25–30 dB) (Richharia 1999), or especially by applying circular polarization.

Inter-system interference may occur between a satellite system and terrestrial system whenever the same frequencies are shared. Certain frequency bands above 1 GHz are shared between the fixed satellite and fixed terrestrial services. Both services are necessary to satisfy the telecommunication needs of the world, so these services have coexisted for over four decades and probably will continue to coexist in the future. With so many telecommunications services using radio transmissions, interference between services can arise in a number of ways. Figure 9.1 shows the possible interference scenarios between satellite and terrestrial services:

- Satellite transmitter interferes into terrestrial station receiver (1)
- Terrestrial station transmitter interferes into satellite receiver (2)

Ground Station Design and Analysis for LEO Satellites: Analytical, Experimental and Simulation Approach, First Edition. Shkelzen Cakaj.
© 2023 The Institute of Electrical and Electronics Engineers, Inc. Published 2023 by John Wiley & Sons, Inc.

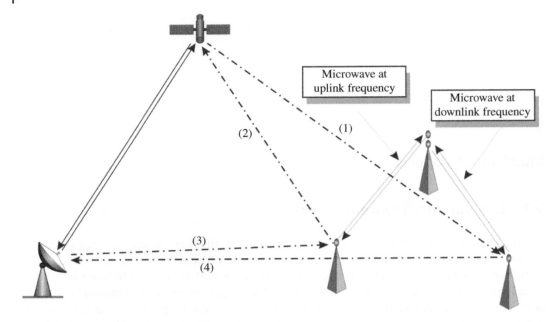

Figure 9.1 Interference scenarios.

- Satellite ground station transmitter interferes into terrestrial station receiver (3)
- Terrestrial station transmitter interferes into satellite ground station receiver (4)

The satellite ground stations typically operate at high elevation angles, while terrestrial stations operate at low elevation angles. This simplifies situation allowing spatial discrimination in both direction (3) and (4) in Figure 9.1. But, for low Earth orbit (LEO) satellites there is a communication under low elevations angles, also, so the interference effects have to be considered. (The term *ground station* is specifically associated with satellite circuits and *terrestrial station* is specifically associated with microwave line of sight circuits.)

Permitted frequency sharing imposes limits on transmission levels for terrestrial transmitters, ground satellite stations, and satellite transmitters in certain bands and services, aiming to minimize intersystem interference. These limitations should be coordinated to reduce interference. Procedures have been developed to permit the coexistence of networks sharing the same frequency by mutual agreement. The procedures and coordination are under International Telecommunication Union (ITU) responsibility. Based on these ITU radio regulations limits, we can apply the following means by which can be reduced the interference for scenarios described in Figure 9.1:

- Limitations on satellite power-flux density (PFD) produced at the surface of the Earth (1).
- Limitations on the terrestrial station EIRP and power delivered to the antenna (2).
- Limitations on the satellite ground station power radiated toward the horizon (3).
- Limitations on distances between satellite ground stations and terrestrial stations (3), (4).
- Antenna performance standards (3).

The values for interfering signals due to frequency reuse cross-polarization, multiple beam interferers, and interference power received from other systems (intra-system and inter-system), must be obtained by carefully constructing the link equation for each case (Difonzo 2000).

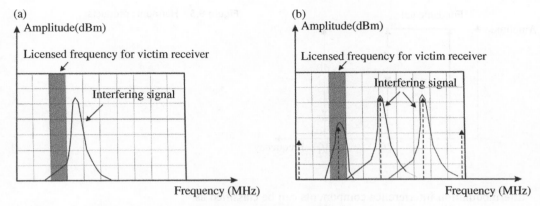

Figure 9.2 Co-channel interference (a), and out-of-band interference (b).

From the technical and practical point of view, two classifications of interference should be considered (Gordon and Morgan 1993). These two scenarios are presented in Figure 9.2:

- Co-channel interference
- Out-of-band interference

The *co-channel interference* occurs when the user's receiver is disturbed by the system or equipment operating at the same frequency as the user's receiver (Figure 9.2a). More problematic is *out-of-band interference*. This interference occurs when the intended receiver is hit by signals that are generated by equipment that does not operate in the same frequency as that receiver. The phenomenon whereby one or more new signals are generated is called *intermodulation*. These new generated signals (*intermodulation products*) can unexpectedly fall within a victim receiver's licensed passband (Figure 9.2b), interfering with signal reception. These unwanted intermodulation products can occur in receivers and may coincide with the operating frequency of the receiver, in which case the wanted signal can be masked. If the signal is too strong, it will completely block the desired receiving signal (Bulloch 1987; Mendenhall 2001; National Telecommunications and Information Systems 2006).

9.2 Intermodulation Products

The influence source of noise in a satellite communication system is the intermodulation noise generated by nonlinear transfer characteristics of devices. Toward the uplink, the intermodulation noise is mainly generated because of the high-power amplifier (HPA) nonlinearity. Related to downlink performance, especially in urban areas (presence of wireless networks, fixed or mobile) intermodulation should be considered because of the low noise amplifier (LNA) nonlinearity. Disturbance introduced due to nonlinearity is known as *intermodulation interference*.

The nonlinear transfer characteristic may be expressed as a Taylor series, which relates input and output voltages:

$$e_0 = ae_i + be_i^2 + ce_i^3 + \dots \tag{9.1}$$

Here, a, b, c, and so on are coefficients, depending on the transfer characteristic, e_0 is the output voltage, and e_i is the input voltage, which consists of the sum of individual carriers.

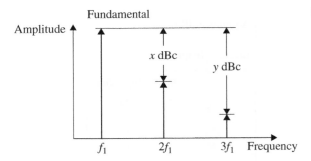

Figure 9.3 Harmonic products.

Intermodulation interference components can be classified as:

- Harmonic products
- Intermodulation products

Harmonic products are single-tone distortion products caused by device nonlinearity. When a nonlinear device is stimulated by a signal at frequency f_1, spurious output signals can be generated at the harmonic frequencies $2f_1$, $3f_1$,... Nf_1. The order of the harmonic products is given by the frequency multiplier; for example, the second harmonic is a second-order product. These harmonics are presented in Figure 9.3. Harmonics are usually measured in dBc (indexed by c), which means dB below the carrier (fundamental) output signal.

Intermodulation products are multi-tone distortion products that result when two or more signals at frequencies $f_1, f_2, ...f_n$ are present at the input of a nonlinear device. The spurious products, which are generated due to the nonlinearity of a device, are related to the original input signals frequencies. Analysis and measurements in practice are most frequently done with two input frequencies (sometimes termed *tones*).

The frequencies of the two-tone intermodulation products are:

$$Mf_1 \pm Nf_2 \text{ where } M,N = ,1,2,3... \tag{9.2}$$

The order of the distortion product is given by the sum $M + N$. The second-order intermodulation products of two signals at f_1 and f_2 would occur at $f_1 + f_2, f_2 - f_1, 2f_1$ and $2f_2$. The third-order intermodulation products (component ce_i^3 of Eq. 9.1) of two signals f_1 and f_2 would be $3f_1, 3f_2, 2f_1 + f_2$, $2f_1 - f_2, f_1 + 2f_2$ and $f_1 - 2f_2$ (Maral and Bousquet 2002; Dodel 1999). These are presented in Figure 9.4. Mathematically, intermodulation product calculation could result in "negative" frequency, but it is the absolute value of these calculations that is of concern. Broadband systems may be affected by all nonlinear distortion products. Narrowband circuits are only susceptible to those in the passband. Bandpass filtering can be an effective way to eliminate most of the undesired products without affecting in band performance (see Figure 9.4).

Third-order intermodulation products are usually too close to the fundamental signals to be filtered out, so the third-order (and to a lesser extent fifth-order) products contribute the major proportion of the intermodulation noise power. The closer the fundamental signals are to each other, the closer third intermodulation products will be to them. Filtering becomes very hard if the intermodulation products fall inside the passband. These unwanted intermodulation products can occur in receivers and may coincide with the operating frequency of the receiver, in which case the wanted signal can be masked. The level of these products is a function of the *power received* and the *linearity of the receiver/preamplifier*. Moreover, the amplitude of the intermodulation product

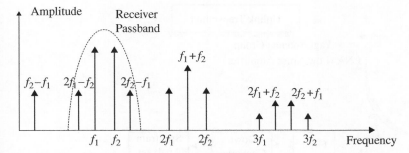

Figure 9.4 Second- and third-order intermodulation products.

decreases with the order of the product. Further, out-of-band intermodulation products transmitted from the ground stations or satellites result in interference to other systems. To minimize such harmful emissions, radio regulations restrict such out-of-band transmissions from ground stations to very low levels. There is no single technical method to eliminate the impact of intermodulation products; on-site experimental investigations are needed (Mendenhall 2001). Such approach and experimental results for LEO ground stations will be further clarified later.

The regulations for various satellite communication and recommendations that affect the planning and design of satellite communication system pertain to:

- Frequency allocations for various satellite communication services
- Constrains on the maximum permissible RF power spectral density from the ground station (CCIR Rec 524)
- Antenna pattern of ground stations (CCIR Rec 465 and 580)
- Constraints on the maximum permissible transmission levels from satellites (CCIR Rec 358)
- Permissible interference from other networks (CCIR Rec 466, Rec 483 and Rec 523)

9.3 Intermodulation by Uplink Signal at LEO Satellite Ground Stations

At the ground station located in an urban area with the high penetration of mobile radio systems, it is not easy to eliminate intermodulation interference signals since these are unpredictable. There is no single technical method to eliminate them, so again, on-site investigations and experimental measurements are needed.

Ground stations in urban areas should be designed so that, at the receiver input, the level of the signal received from the satellite via the main beam of the ground station antenna exceeds the in-band noise by an adequate margin. But, the unwanted out-of-band inputs, as intermodulation products, generated by the ground station uplink signal and signals from nearby mobile system base stations, even though they are received via sidelobes in the ground station's antenna pattern, they could be higher and can mask the wanted signal (Maral and Bousquet, 2002; Sklar 2001; Mendenhall 2001).

Idea: The intention is to mathematically and by executed experiment confirm whether the intermodulation products will disturb the receiver at the Vienna satellite ground station dedicated for communication with MOST satellite.

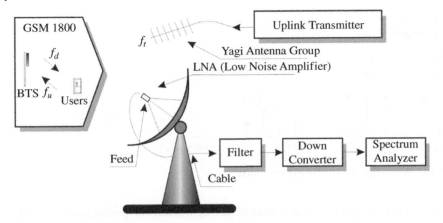

Figure 9.5 Intermodulation scenario at satellite ground station.

Method: Math analysis and experimental execution is applied. Figure 9.5 presents the experiment setup that enables us to generally check the intermodulation disturbance at the any receiving satellite ground station (Cakaj et al. 2005).

In Figure 9.5 in front of satellite receiving ground station the GSM 1800 signals are presented. The similar procedure could be used in case of other MW signal presence, also. As in case of intermodulation, each specific case specifically should be studied. Thus, here we analyze further the intermodulation disturbance at Vienna satellite ground station within MOST scientific space observation project.

The presence of intermodulation products, at ground station, near the downlink frequency ($f_r = 2232$ MHz) caused by GSM 1800 and uplink signal ($f_t = 2055$ MHz) is expected because of eventual nonlinearity of the LNA used in the front end at the downlink of the ground station. By the nonlinearity of the LNA, the intermodulation products will be generated from the uplink signal at frequency f_t on one hand and GSM signals at frequencies f_{GSM} on the other. Only third-order intermodulation products will be considered.

In order to make correct analysis, www.rtr.at provides the GSM frequency plan related to GSM 1800 providers in Austria. These data are presented in Table 9.1 (Cakaj et al. 2005), showing the first channel, a few in the middle, and the last one.

Table 9.1 Frequencies of GSM 1800 providers operating in Austria.

Channel	f_u	f_d	Provider
512	1710.2 MHz	1805.2 MHz	TMA
521	1712.0 MHz	1807.0 MHz	TMA
523	1712.4 MHz	1807.4 MHz	Mobilkom
586	1725.0 MHz	1820.0 MHz	Telering
632	1734.2 MHz	1829.2 MHz	One
868	1781.4 MHz	1876.4 MHz	One

In Table 9.1 f_u is the GSM uplink signal frequency and f_d is the GSM downlink signal frequency of the GSM 1800 network (see Figure 9.5). Based on ITU-R F.382-6, 1.7GHz–2.1GHz frequency band for mobile systems is 1710 MHz–1785 MHz for the uplink and 1805 MHz–1880 MHz for the downlink. Recall that the difference between the upper edge of the band and the last frequency within a band is called *guard band* (GB). So, in this case the guard bands are:

$$GB_u = 1785\,\text{MHz} - 1781.4\,\text{MHz} = 3.6\,\text{MHz} \tag{9.3}$$

$$GB_d = 1880\,\text{MHz} - 1876.4\,\text{MHz} = 3.6\,\text{MHz}$$

Signals present at the front end of the preamplifier of the receiving system at the ground station are presented in Figure 9.6. Intermodulation products generated by signals at frequencies f_t and f_u fall too far on the frequency domain from the receiver's downlink frequency f_r; therefore, they will not be considered here. Third-order intermodulation products generated by frequencies f_t and f_d are $2f_t \pm f_d$ and $2f_d \pm f_t$.

Only, products $2f_t - f_d$ are worth further analysis, because only they fall in the frequency domain near the receiver's frequency f_r. These intermodulation products appear at the preamplifier's (LNA) output (respectively at the filters input) in frequency domain (RF) (see Figure 9.7) (Cakaj et al. 2005; Cakaj and Malaric 2007).

The respective frequencies of these signals correlated to Table 9.1 are presented in Table 9.2.

These signals will be filtered before going into the downconverter (see Figure 9.5). The situation behind the filter and in front of the downconverter is presented in Figure 9.8.

From Figure 9.8 it is clear that the filter has substantially attenuated a considerable number of interference contributions from intermodulation products. The local oscillator frequency of the

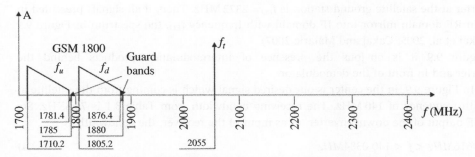

Figure 9.6 Signals present at frontend of preamplifier (LNA) of the downlink.

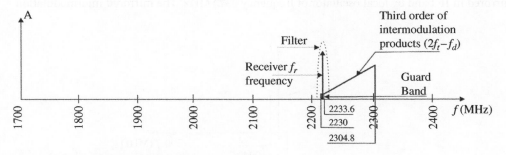

Figure 9.7 Third order of intermodulation products.

Table 9.2 Third order intermodulation products.

f_t	f_d	$2f_t - f_d$
2055 MHz	1805.2 MHz	2304.8 MHz
2055 MHz	1807.0 MHz	2293.0 MHz
2055 MHz	1807.4 MHz	2292.6 MHz
2055 MHz	1820.0 MHz	2290.0 MHz
2055 MHz	1829.2 MHz	2280.8 MHz
2055 MHz	1876.4 MHz	2233.6 MHz

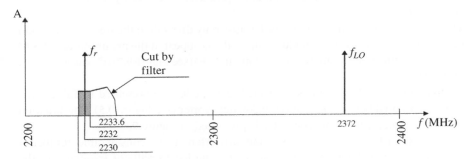

Figure 9.8 Signals in front of downconverter.

downconverter at the satellite ground station is $f_{LO}=$ 2372 MHz. Then, if all signals presented in Figure 9.8 in RF domain mirror into IF domain with frequency f_{LO}, the spectrum in Figure 9.9 follows (Cakaj et al. 2005; Cakaj and Malaric 2007).

From Figure 9.9 it is obvious the presence of intermodulation products behind the downconverter and in front of the demodulator.

Results: In Figure 9.9, in the center is our desired signal which is coming from the satellite at IF input with frequency of 140 MHz. The receiving bandwidth from Table 3.1 is 76.8KHz. So, looking at IF output of the downconverter, or as input of the receiver, the bandwidth is

$$139.9616MHz < f < 140.0384MHz \qquad (9.4)$$

The intermodulation interference will affect the desired signal if respective intermodulation frequencies fall within this range. From Table 9.2 the intermodulation products (third column) are mirrored in IF band by local oscillator of frequency 2372 MHz. The mirrored intermodulation

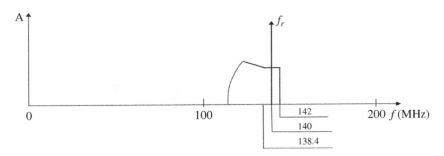

Figure 9.9 Downconverter output.

Figure 9.10 Presence of intermodulation products.

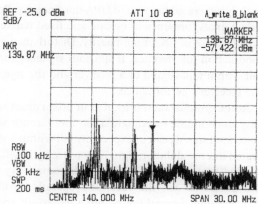

products are: 67.2, 79, 79.4, 82, 91.2, and 138.4 MHz. No one of these intermodulation products is within a frequency range under Eq. (9.4), so there is no intermodulation interference. In this case, an advantage is that satellite downlink frequencies lie in a band that is in fact the mirrored guard band.

Intermodulation products are recorded by experiment, and the appropriate set up is presented in Figure 9.5. The IF record by spectrum analyzer confirming a mathematical calculation is presented in Figure 9.10. (Cakaj et al. 2005; Cakaj and Malaric 2007).

In the center of Figure 9.10 is the desired signal coming from the satellite. On the left side there are intermodulation products that do not mask the desired receiving signal. The span in Figure 9.10 is 30 MHz. The receiving bandwidth from Table 3.1 is 76.8 KHz, so although intermodulation products are present, they are too far from the receiving bandwidth to significantly degrade the signal. The advantage of this case is that the satellite downlink frequencies lie in a band that is in fact the mirrored guard band of GSM 1800 MHz in IF domain (Figure 9.8 and Figure 9.9). The envelope around signal f_r at Figure 9.9 stemmed from mathematical analysis and the envelope of the experimentally recorded signal around f_r are almost identical, in this case confirming each other. This makes possible to have LNAs directly connected to the feed without filter, and therefore the *maximum downlink margin*.

Conclusion: The final conclusion is that the presence of the 50 W uplink signal and the GSM 1800 signals do not produce intermodulation products that disturb the performance of the down-link receiving system at the (LEO) satellite ground station. This makes it possible to have LNAs directly connected to the feed without filter, and therefore the maximal downlink margin. But, in case of the intermodulation disturbance, the filter should be implemented in the front end, more exactly in between the feeder and LNA.

9.4 Modeling of Interference Caused by Uplink Signal for LEO Satellite Ground Stations

Idea: Interference can negatively affect the functionality of the satellite ground station. The major source of interference in a satellite communication system is intermodulation noise generated by nonlinear transfer characteristics of devices. Toward the uplink, the intermodulation noise is mainly generated because of HPA nonlinearity. Related to the downlink performance, especially in urban areas (presence of mobile and fixed wireless networks) intermodulation should be considered because of LNA nonlinearity. Disturbance introduced due to nonlinearity is known as intermodulation interference. If spurious signals generated as intermodulation products behind LNA fall within a passband of a receiver and the signal level is of sufficient amplitude, it can degrade the receiver's performance.

At the ground stations located in urban areas with high density of mobile radio systems it is not easy to eliminate intermodulation interference signals, since these are unpredictable. These new generated signals can unexpectedly fall within a victim receiver licensed passband. In case the generated intermodulation signal is too strong, it will not only interfere but could completely block the desired receiving signal. Filtering becomes very hard if the intermodulation products fall inside the passband. So, the receiver's operation will be disturbed if two conditions are fulfilled: (i) interference signal fall within a passband; and (ii) has too high power.

Based on this concept and the example described under Section 9.3, the idea is to build the intermodulation interference modeling flowchart that enables the interference calculation caused by any other radio source of frequency f_x and the satellite uplink signal of frequency f_t. From the appropriate model will be built the interference calculator to determine and predict such disturbances of the receiver, based on the radio signals environmental presence.

Method: Based on interference concepts and the previously mathematical interpretation, considering radiofrequency signal (f_x) present in the front end of the satellite ground station's receiving system, which is potential for generating intermodulation interference, it is analyzed and then modeled the intermodulation interference, and further is introduced the intermodulation interference calculator.

Only third-order intermodulation products are considered. Among third-order intermodulation products are considered only components of frequencies $2f_x - f_t$ and $2f_t - f_x$. Analyses under Section 9.2 confirmed that these products could fall within a receiver's passband. Other intermodulation products of frequencies $3f_x$, $3f_t$, $2f_x + f_t$ and $2f_t + f_x$ usually fall too far from the passband and practically are eliminated by filtering. Thus, these products are not treated in the modeling approach (Cakaj et al. 2008; Cakaj 2010).

The amplitudes of intermodulation products of frequencies $2f_x - f_t$ and $2f_t - f_x$ are respectively $3A_x^2 A_t$ and $3A_t^2 A_x$ (these yields out from trigonometry) where A_x is amplitude of any radio signal of frequency f_x in front of LNA, which has the potential to cause intermodulation with uplink satellite signal of frequency f_t and amplitude A_t. Thus, third-order intermodulation products are characterized by:

$$f_{i1} = 2f_x - f_t, N_{i1} = 3A_x^2 A_t \tag{9.5}$$

$$f_{i2} = 2f_t - f_x, N_{i2} = 3A_t^2 A_x \tag{9.6}$$

where f_{in} is intermodulation interference frequency of amplitude N_{in} for $n = 1, 2$ behind the LNA. Since, the analyses are related mainly to the frequency domain, in order to simplify the situation, it is assumed that there is no amplification on overall system chain.

Usually, the amplitude A_x is low in front of LNA since it is limited by ITU rules about radiated power and consequently it is expected that the amplitude $N_{i1} = 3A_x^2 A_t$ will not disturb the receiver. The most dangerous component is $N_{i2} = 3A_t^2 A_x$ since the amplitude A_t is high because this is the amplitude of uplink signal that must overcome too-high attenuation toward the satellite.

The reference checking point is downconverter's IF output or receiver's IF input. So, the intermodulation interference is checked around intermediate frequency f_{IF}. The mirroring into intermediate frequency is achieved by local oscillator frequency of f_{LO}. All frequencies are mirrored by f_{LO}, including intermodulation products and desired receiving signal of frequency f_r. Thus, it is:

$$f_{IF} = f_{LO} - f_r \tag{9.7}$$

For receiver with bandwidth $B = 2\Delta f$, the receiving passband at IF input is from $f_{IF} - \Delta f$ up to $f_{IF} + \Delta f$ where f_{IF} is the intermediate frequency, which is usually 140 MHz or 70 MHz. Thus, the receiver could be disturbed if the intermodulation product mirrored at IF falls within frequency band at IF input, mathematically expressed as:

$$f_{IF} - \Delta f \leq f_{in} - f_{LO} \leq f_{IF} + \Delta f \tag{9.8}$$

By substituting f_{IF} from Eq. (9.7) to Eq. (9.8) yields

$$(f_{LO} - f_r) - \Delta f \leq f_{in} - f_{LO} \leq (f_{LO} - f_r) + \Delta f \tag{9.9}$$

Then, further, if we substitute f_{in} from Eq. (9.5) and Eq. (9.6) at Eq. (9.9) will have:

$$(f_{LO} - f_r) - \Delta f \leq (2f_x - f_t) - f_{LO} \leq (f_{LO} - f_r) + \Delta f \tag{9.10}$$

$$(f_{LO} - f_r) - \Delta f \leq (2f_t - f_x) - f_{LO} \leq (f_{LO} - f_r) + \Delta f \tag{9.11}$$

Thus, if intermodulation frequency fulfills the Eq. (9.10) or Eq. (9.11) the desired signal at the receiver could be masked by intermodulation interference.

The next step is considering power (amplitude) issue. Thus, the level of interfering signal should be compared with the level of desired signal at IF input. For comparison of these levels, it is sufficient to consider the relationship between the two relative levels (one of them is the reference level). Usually this is measured with a spectrum analyzer at IF checkpoint. The criteria for amplitudes comparison between the desired and interference signal depends on the Earth's station size and dedication. The criteria between downlink carrier level and interference signal level range from 20 to 30 dB (http://www.satsig.net/interfer.html). This is mathematically expressed by Eq. (9.12):

$$S_{(IF)}(dB) - N_{in(IF)}(dB) \geq (20 \div 30)dB \tag{9.12}$$

where $S_{(IF)}$ is desired signal power and $N_{in(IF)}$ intermodulation interference signal power at IF input. These two power levels can be calculated or measured in order to conclude about the receiver's disturbance. The above concept is presented through the flowchart in Figure 9.11. Input parameters in Figure 9.11 are: f_x is frequency of radio source in front of LNA of the satellite receiving system, f_t uplink transmit frequency, f_r downlink receiving frequency, and B is downlink receiver's bandwidth. Considering the appropriate flowchart, it is structured the following intermodulation interference calculator presented in Figure 9.12 (Cakaj et al. 2008; Cakaj 2010).

Usually, only one of the treated components falls within a passband and causes the disturbance. So, if intermodulation components are out of band, then under status for f_{in}, $n = 1, 2$ will show up this text: "Intermodulation products out of band (no disturbance)," and no further analyses are needed.

In case when one of components falls inside the band, then under status for f_{in}, $n = 1, 2$ will show up this text: "In band interference" and further analyses related to the amplitude level are needed. If amplitude of interference is under limited level at final status will show up text as "No disturbance," and if the amplitude of interference level is above planned limit shows up the text: "It is disturbance (Action)."

Results: This study is further confirmed under a particular case, where uplink transmits with frequency $f_t = 2055$ MHz, and in the surrounding environment are present GSM 1800 signals designated as f_d. The third-order generated intermodulation products frequencies are $(2f_t - f_d)$, as given

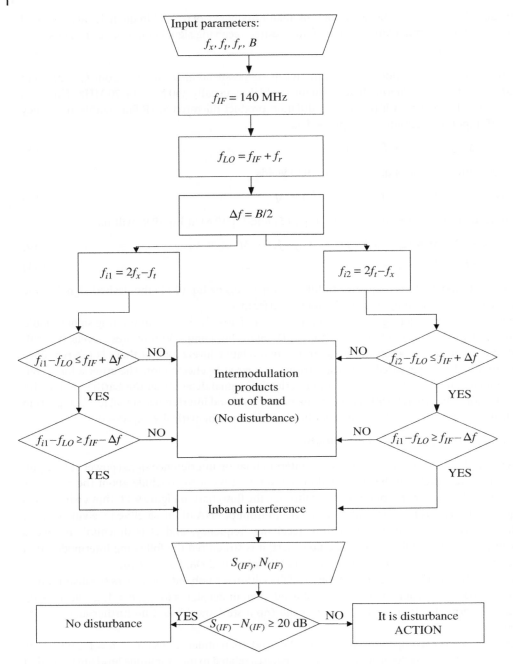

Figure 9.11 Intermodulation interference modeling flowchart.

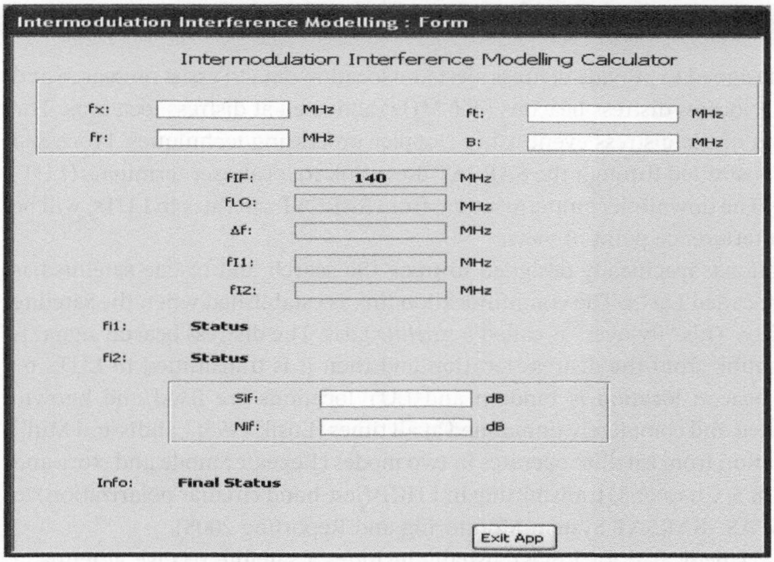

Figure 9.12   Intermodulation interference calculator.

in Table 9.2, Section 9.2 (Cakaj et al. 2005). Further, for $B = 100$KHz, $f_{IF} = 140$ MHz and $f_r = 2232$ MHz, $f_{LO} = 2372$ MHz, the receiving bandwidth, at IF output of the downconverter or as input of the receiver, is:

$$139.95 MHz < f < 140.05 MHz \tag{9.13}$$

f_{IF}, f_r, f_{LO} are, respectively, intermediate, receiving, and local oscillator frequencies. From Table 9.2 the intermodulation products (third column) are mirrored in IF band by local oscillator of frequency 2372 MHz. The mirrored intermodulation products are: 67.2, 79, 79.4, 82, 91.2, and 138.4 MHz. No one of them falls within mirrored IF bandwidth, thus the conclusion is that there is no downlink disturbance by intermodulation products. Finally, for the case presented under Section 9.3, at Figure 9.5 the intermodulation interference disturbs receiving system if in front of LNA is present a signal of frequency $f_x = 1598$ MHz or $f_x = 2283.5$ MHz. This is confirmed by applying intermodulation interference modeling.

Conclusion: The introduced modeling concept could be applied on uplink signal frequency selection in order to avoid the interference. These analyses are of high importance on the final decision of the ground station design. This methodology is applicable for medium Earth orbit (MEO) MEO systems, also. The math analysis and experimental results, under 9.3, are completely in accordance with results stemmed from the modeling approach.

9.5   Downlink Adjacent Interference for LEO Satellites

The downlink adjacent interference is expected when two satellites operate in close proximity to each other and share the same frequency, affecting the desired signal-to-noise ratio of the downlink. Further, this problem is considered and analyzed under very practical operational circumstances for SARSAT (Search and Rescue Satellite Aided Tracking) system (constellation).

The SARSAT system is operated and managed by NOAA (National Oceanic and Atmospheric Administration).

The SARSAT system is designed to provide distress alert and location data to assist on search and rescue operations. SARSAT locates distress beacons (406 MHz) activated at distress locations. The system calculates a location of the distress event using Doppler processing techniques. Processed data are continuously retransmitted through the SARSAT downlink to local user terminals (LUT) when satellites are in view. The downlink communication, from SARSAT satellites to LUTs, will be further treated from the interference point of view.

Receive-only ground stations, specifically designed to track the search and rescue satellites as they pass across the sky are called LUTs. The communication link is established when the satellite flies within a LUT's visibility. This "fly-over" is called a *satellite pass*. The distress beacon signal is received on the satellite uplink from the distress location and then it is transmitted to LUTs by downlink. Normally, the beacon location is random and LUT locations are fixed and known. The LUTs are fully automated and completely unmanned at all times [Losik 1995; Landis and Mulldolland 1993]. Communication from satellite operates in two modes (Repeater mode and store and forward mode, Figures 8.4, 8.5, Chapter 8) transmitting in LHCP (left-hand circular polarization) to any LUT in its view (COSPAS –SARSAT System Monitoring and Reporting 2008).

The LEOLUT (means LUT dedicated for LEOs), usually includes a satellite receive antenna, a digital processing system, and the software for control, monitoring and processing functions. When a satellite receives a beacon signal from a distress location, the Search and Rescue Processor (SARP) on board the LEO satellite performs Doppler processing and generates an entry into the 2.4 kb/s processed data stream (*pds*) that is continuously "dumped" to any LEOLUT in view of the satellite's downlink footprint. (COSPAS –SARSAT System Monitoring and Reporting 2008).

LEOLUT software accepts the satellite's downlink data stream, then decodes and extracts beacon data messages. From each satellite pass taken by the LEOLUT, software selects data from each detected beacon and validates time, frequency (Doppler shifted frequency), and message content. Data from each pass, and for each beacon identification number, is then passed to the solution processing software. The solution processing software determines an optimum location based on a Doppler frequency curve, built based on frequency shift. An example of recorded Doppler curve at LUT is given in Figure 9.13.

This is an excellent Doppler curve providing 15 Doppler events, noted from 1 to 15. Time is given on the *x*-axis and Doppler frequency shift on the *y*-axis. Using the orbital parameters of the satellite, the beacon frequency, and the known Doppler shift, the distance of the beacon relative to the projection of the satellite orbit ground-track on the Earth can be determined as is depicted in Figure 8.10 (Vataralo et al. 1995; Cakaj et al. 2010a).

Communication reliability during a satellite pass may be degraded when satellites sharing the same downlink frequency are adjacent to each other. The downlinks of all SARSAT LEO satellites use the same 1544.5 MHz frequency (COSPAS – SARSAT 406 MHz Frequency Management Plan, 2008). In cases where the satellites are within the main lobe of the local user terminal antenna, transmissions from adjacent satellites act as interference to one another, consequently degrading the desired signal at the appropriate LUT (Vataralo et al. 1995; Cakaj and Malaric 2007). This can result in missed distress beacon bursts or no stored solutions received at the LUT, consequently no data is provided about a distress location. This should be avoided as much as possible!

Analysis on interference prediction, interference periods, and mitigating procedure are further discussed. Interference mitigation of significant duration, with attached measurement results, is also presented. For further analytical purposes, to avoid repetition, we use the data from Table 8.1 and Table 8.2, which present the space and ground segment of the SARSAT system.

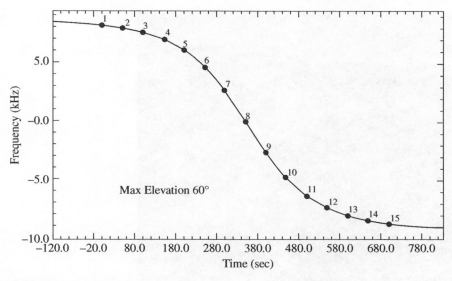

Figure 9.13 Doppler curve.

9.6 Adjacent Satellites Interference (Identification/Avoiding)

Satellites are considered as adjacent if they are too close to each other determined by their space orbital parameters, or too close in radar map or as the ground tracks, which for further analysis is applied. The adjacent satellite interference manifests when two adjacent satellites share the same downlink frequency, and are seen too close to each other from the ground station. If the transmitted EIRP (equivalent isotropic radiated power) from each satellite is similar, for two satellites close to each other, the two signals will act as interference to each other, severely degrading the received desired signal (Kanellopoulos et al. 2007; Cakaj et al. 2010a). The above is illustrated in Figure 9.14 where the receiving ground stations noted as AK, CA, and HI (as a part of SARSAT ground segment) are looking at two satellites S9 and S11(a part of SARSAT space segment).

The downlink of all SARSAT LEO satellites uses the same 1544.5 MHz frequency, and in principle the same transmitted EIRP from each satellite. Thus, if two satellites are too close to each other, their signals will interfere with each other, severely degrading the received signal (COSPAS – SARSAT 406 MHz Frequency Management Plan, 2008). The received carrier frequency provides a useful measure of the interference level. The carrier frequency of the transmitter is 1544.5 MHz, but the relative velocity between the satellite and LUT causes a Doppler shift in the received frequency, and a plot over time shows the characteristic Doppler curve of a LEO satellite. As the orbital positions of the two satellites converge, so do their relative velocities to the LUT and Doppler curves (COSPAS –SARSAT System Monitoring and Reporting 2008).

Slight differences in relative velocity between the two interfering satellites cause two distinct curves of carrier frequency. When the difference in relative velocity and angular separation is minimal, the Doppler curves of the carrier frequency become almost identical. Figure 9.15 shows the real-time received carrier frequency for Florida-1 and Florida-2 LUTs. Florida-1 is tracking S9 and Florida-2 is tracking S11 (Cakaj et al. 2010a).

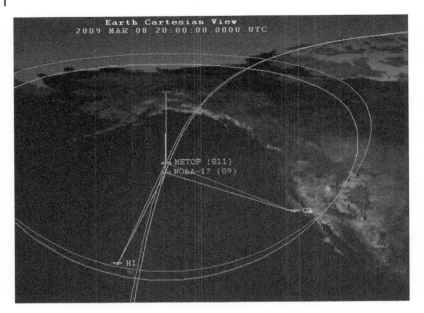

Figure 9.14 Adjacent satellites seen from the ground station.

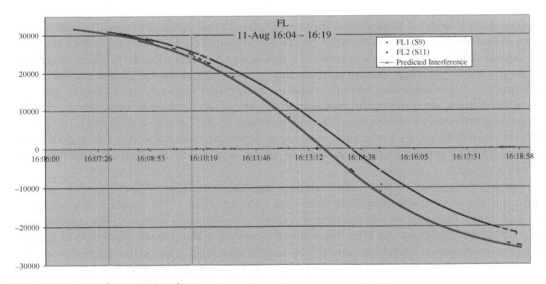

Figure 9.15 Interference interval.

When the receiver locks on the interfering signal, jumping (interruptions on curve) of the received carrier frequency is seen. These interruptions in carrier lock result in loss of downlink capability and can visually show when interference has occurred. Figure 9.15 shows that for most of the pass, each LUT is successfully locked on its desired signal. Two vertical lines, in Figure 9.15 show the time interval of interference. This can result in missed bursts (missed Doppler event), or no signal received at all.

Table 9.3 Passes affected by interference.

Date	DOY	AOS	LOS	LUT	SAT	Orbit	Reason
03.08.09	067	16 : 31	16 : 43	LSE	S9	34 844	No *pds* solution
03.08.09	067	16 : 35	16 : 47	CA2	S9	34 844	No *pds* solution
03.08.09	067	18 : 08	18 : 18	AK2	S9	34 845	No *pds* solution
03.08.09	067	19 : 56	20 : 04	CA1	S9	34 846	No *pds* solution
04.16.09	106	19 : 49	19 : 58	CA2	S9	35 401	No *pds* solution
04.16.09	106	21 : 34	21 : 44	HI2	S9	35 402	No *pds* solution
04.17.09	107	0 : 58	1 : 11	GU2	S11	12 932	No *pds* solution

For SARSAT system, the downlink interference between S11 and S9 (Figure 9.15) was documented by France in March and April 2009, when S9 and S11 were close to each other. The March 8, 2009, occurrence of interference between these two satellites caused four passes, over a period of three orbits, which produced no *pds* solutions. The April 16, 2009, occurrence of interference caused three passes with no *pds* solutions over a period of three orbits, presented in Table 9.3 (DOY means day of the year). But, the number of no *pds* solutions alone cannot accurately gauge the amount of interference in the downlink. It is a significant variability in the number *pds* bursts received by the satellite during each orbit depending on the path.

Considering orbital parameters (Table 8.1), three pairs of operational SARSAT satellites are susceptible to this interference condition: S10/S12, S9/S11, and S7/S8 identified and presented in Table 9.4 with their respective orbital periods and differences between them, and recorder by NOAA orbital software are given in Figure 9.16.

The small difference in orbital periods of the S10/S12 pair is particularly concerning. The *Orbit repeat cycle* indicates the number of orbits that satellite should pass through to achieve the same position relative to the adjacent satellite and to the fixed ground station. Mathematically, *Orbit repeat cycle* is the ratio of orbit period and orbital difference. Further, for this cycle to be expressed in days, it should be divided by the mean motion from Table 8.1 (for example for S12 is 14.1095). The SARSAT documented that the launch of S12 (NOAA-19) into an orbital plane similar to S10 (NOAA-18), and with nearly identical orbital periods, created long *periods of adjacent interference*. The first period of extended interference occurred September 15–20, 2009 (Cakaj et al. 2010a).

Table 9.4 SARSAT adjacent satellites.

Satellite	Orbit period	Difference	Orbit repeat cycle	Repeat cycle (days)
SARSAT12	01 : 42 : 03.53	00 : 00 : 01.30	4710	334
SARSAT10	01 : 42 : 02.23			
SARSAT11	01 : 41 : 18.10	00 : 00 : 10.92	556	39
SARSAT 9	01 : 41 : 07.18			
SARSAT 8	01 : 41 : 56.75	00 : 00 : 52.55	116	8
SARSAT 7	01 : 41 : 04.20			

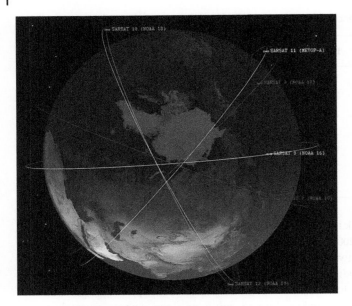

Figure 9.16 SARSAT adjacent satellites.

9.6.1 Adjacent Interference Identification and Duration Interval

To determine the duration of the interference periods, one must find the minimal angular separation between satellites as seen from the ground station, when interference occurs. This is highly dependent on the gain pattern and pointing accuracy of the LUT antenna. For a typical LEOLUT antenna gain pattern, the −3 dB (half power) beamwidth is found to be ±4.25°. This beamwidth represents the necessary angular separation to prevent undesired signals from being highly amplified. As the angular separation increases, the gain of the interfering source decreases. Since the distance between the two satellites is relatively constant during a singular pass, it can be seen that the apparent angular separation is greatest when the satellites are at their maximum elevation (closest approach) (Cakaj et al. 2010a). Thus, minimum angular separation occurs when the satellites are at minimum elevation. Thus, the cases with low elevation are of interest from the interference aspect.

Idea: Under the case that the satellites based on their orbital parameters are adjacent, and the interference is expected, the main question is to conclude how long it will take? Further elaborated!

Method: Mathematical and simulation approach are applied and confirming each other. Let us consider a LUT with antenna aperture of ±4.25°. This antenna is tracking a satellite, which is moving ahead relative to another satellite which is seen at minimum elevation above the horizon (5°), as shown in Figure 9.17.

These adjacent satellites, seen at low elevation and with a very low separation angle, have great potential to interfere with each other. The slant range is calculated for elevations of 9.25° and 5° (5° is designed horizon with 4.25° separation) from the ground station. Spatially the separation angle is the spherical angle from 0° to 4.25°. The 0° point is on the desired satellite, and 4.25° point is the −3 dB interference point; consequently, it is the largest possible distance for interference from another satellite. The general formula for the slant range (*d*) under elevation ε_0 is (Cakaj and Malaric 2007):

$$d = R_E \left[\sqrt{\left(\frac{H + R_E}{R_E}\right)^2 - \cos^2\varepsilon_0} - \sin\varepsilon_0 \right]$$

(9.14)

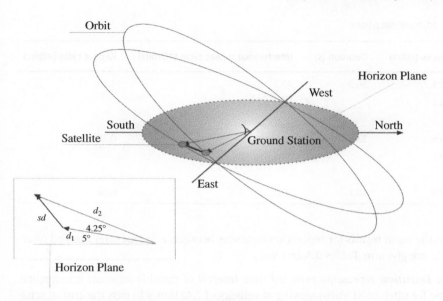

Figure 9.17 Adjacent satellites angular separation.

where, $R_E = 6378$km is Earth radius and H is orbital altitude. The separation distance (sd) can then be determined using a small angle approximation and applying cosines theorem, as:

$$sd = \sqrt{d_1^2 + d_2^2 - 2d_1d_2 \cos 4.25°} \qquad (9.15)$$

Where d_2 is the slant range of pointed satellite from the ground station and d_1 is the slant range of the adjacent satellite potential to interfere.

Altitude H for each satellite is taken from Table 8.1. The slant ranges and separation distances calculated based on Eq. (9.14) and Eq. (9.15) are presented in Table 9.5. For more exact calculations, these separation distances should be multiplied by cosines of separation angle (projection of separation distance in its own orbit), which for too-low angles can be considered as 1. This separation distance when interference may occur, and the difference in orbital periods can then be used to find the duration of possible interference. Considering that these satellites are always moving with a particular velocity v, the question is how long they can be together within a separation angle of 4.25° or lower!

Table 9.5 SARSAT adjacent satellites.

Satellite	Slant range (at 9.25°) (km)	Slant range (at 5°) (km)	Separation distance (km)
SARSAT12	2544.5		416.4
SARSAT10		2903.8	
SARSAT11	2470.3		399.1
SARSAT 9		2812.9	
SARSAT 8	2534.4		341.4
SARSAT 7		2806.5	

Table 9.6 SARSAT adjacent satellites.

Satellite	Velocity (km/s)	Duration (s)	Interference repeat cycle (#Orbits)	Repeat cycle (#days)
SARSAT12	7.423	56.3	43.3	3.10
SARSAT10	7.423			
SARSAT11	7.446	53.9	4.9	0.35
SARSAT 9	7.441			
SARSAT 8	7.447	46.1	0.9	0.06

Results: Finally, the math results for separation distances between adjacent satellites and interference repeat cycle are given in Tables 9.5 and 9.6.

In Table 9.6, the ***Duration*** represents expected time interval of possible adjacent interference. (This is needed time for satellite S12(S10) moving at velocity of 7.423km/s to pass the critical separation distance of 416.4km). The frequency of these events and their duration relative to the fixed ground stations depend on the difference of orbital period times (Table 9.4). The ratio of interference time duration to time difference in orbital periods represents *Interference repeat cycle per orbit*. For duration of 56.3s divided by orbital time difference of 1.3s from Table 9.4 the repeat cycle is 43.3. This cycle is expressed in days when divided by mean motion (Table 8.1, it is 14.1095). Thus, for repeat cycle of 43.3 divided by 14.1095 stems the value of approx. 3.10 days. Considering separation distance, predictions for the interference repeat cycle of satellite pairs are listed in Table 9.6. Math results from Table 9.6 are in full accordance with outputs from orbital software given in Table 9.4.

Conclusion: In general, the difference in orbital periods between the two satellites will dictate the duration and repeatability of interference intervals. From Table 9.6 it is obvious that the S10/S12 pair experiences the highest interference repeat cycle – consequently, the longest possible interference disturbance, because of too-close orbital periods. This means that the satellite pair S10/S12 is of particular concern from the interference aspect. The S7/S8 pair is the least experiences with interference (obvious from Figure 9.16).

9.7 Modulation Index Application for Downlink Interference Identification

On support of interference identification, LEOLUT's software uses two more parameters: modulation index (mean) and modulation index (root mean square – RMS). Modulation index indicates the quantity by which the modulated variable varies around its unmodulated level (Cakaj 2012). Considering downlink phase modulation, modulation index (mean) relates to the variations in the phase of the carrier signal, expressed as:

$$m_i = \Delta\theta(t) \tag{9.16}$$

$\Delta\theta(t)$ is the phase deviation.

The other measure is modulation index (root mean square). In mathematics, the root mean square (RMS) is a statistical measure of the magnitude of varying quantity. It is especially useful for sinusoids wave forms. In case of set of n values m_{i1}, m_{i2}, ...m_{in}, the RMS is given by (Mean-Root-Square, 2012; Yafeng et al. 2004):

$$m_{RMS} = \sqrt{\frac{m_{i1}^2 + m_{i2}^2 + ... + m_{in}^2}{n}}$$ (9.17)

The corresponding formula for a continuous function $m(t)$ defined over the time interval $T_1 < t < T_2$, the RMS is:

$$m_{RMS} = \sqrt{\frac{1}{T_2 - T_1} \int_{T_1}^{T_2} |m(t)|^2 dt}$$ (9.18)

For illustration and interpretation, two different Doppler curves are given, with no interference and the next one with interference presence, presented in Figures 9.18 and 9.19, respectively.

The received carrier frequency is shown on the left axis, and modulation index mean and RMS (route mean square) on the right axis. Modulation index indicates the quantity by how much the modulated variable varies around its unmodulated level. Considering downlink phase modulation, this index relates to the variations in the phase of the carrier signal.

In Figure 9.18, there is no interference, so the lines expressing modulation index and root mean square index are almost flat, confirming noninterference presence. Figure 9.19, is the case with medium maximal elevation of 36°. Figure 9.19 shows the interference during AOS (acquisition of satellite- left circled part). Further, as the satellite moves toward higher elevation there is no interference (medium part of figure) and then again there is interference near LOS (loss of satellite-right circled part). Frequency jumps in the downlink carrier can be seen in the lower-right corner of Figure 9.19, manifested by a high mean modulation index at the same time. In Figure 9.19, it is very expressive modulation index and frequency jump during the loss of satellite. In Figure 9.19 modulation index is circled and mean root square index is squared, confirming exactly the principles discussed.

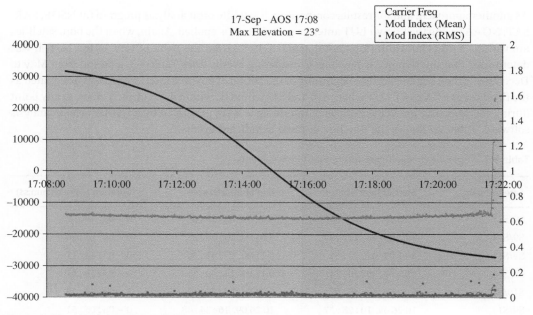

Figure 9.18 No interference Doppler curve for Max. El. 23 °.

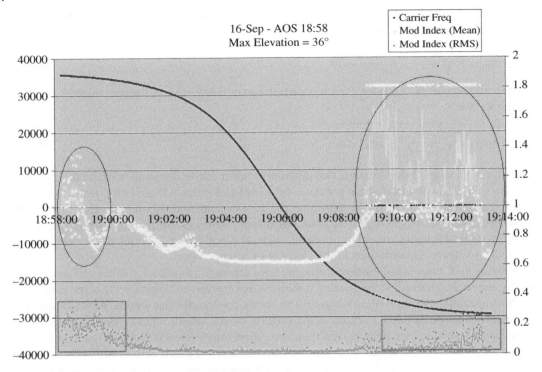

Figure 9.19 Doppler curve for Maximal elevation of 36° (present interference).

9.7.1 Simulation Approach of Interference Events and Timelines

As another approach, and for results comparison, a satellite orbit analysis program (at NSOF, SAR-SAT, NOAA) using the known LUT antenna gain pattern is applied. Again, when the both satellites are within –3 dB beamwidth (separation angle of 4.25°) from the point of view of the LUT, it was determined that interference is possible. Considering events from Table 9.3, a period from May to December is then conducted. The beginning of the period of possible interference was designated as the first pass at a SARSAT LUT where S10 and S12 would be within 4.25° of each other, at any point during the pass. The predicted periods of interference were generated by a NOAA orbital analysis software. Table 9.7 shows the timeline of these significant events.

Table 9.7 Timeline of significant future interference events.

Satellite pair	Start of interference	End of interference	Duration (days-hh:mm:ss)
S9/S11	5.25.09, 18 : 06 : 59	5.26.09, 02 : 26 : 06	0 – 08 : 19 : 07
S9/S11	7.03.09, 13 : 11 : 29	7.03.09, 19 : 37 : 15	0 – 06 : 25 : 46
S9/S11	8.11.09, 01 : 29 : 55	8.11.09, 07 : 41 : 06	0 – 06 : 11 : 11
S9/S11	9.18.09, 09 : 20 : 34	9.18.09, 14 : 40 : 31	0 – 05 : 19 : 57
S10/S12	9.20.09, 10 : 49 : 37	9.23.09, 19 : 54 : 12	3 – 09 : 04 : 35
S9/S11	10.26.09, 10 : 28 : 57	10.26.09, 16 : 34 : 48	0 – 06 : 05 : 51
S9/S11	12.03.09, 07 : 23 : 15	12.03.09, 12 : 37 : 25	0 – 05 : 14 : 10

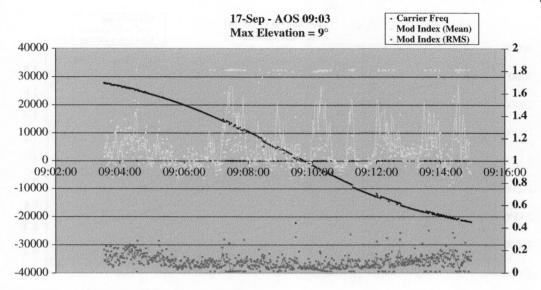

Figure 9.20 Doppler curve for Maximal elevation of 9°.

The appropriate software confirmed that the duration of the S10/S12 interference is of particular concern, and has been verified through this simulation method to be about three days. This proves the accordance between mathematical analysis results (Table 9.6) and simulation results given in Table 9.7 (Cakaj et al. 2010a). The duration of S9/S11 interference periods decreases. Periods of S8/S7 interference are negligible and therefore are not listed.

Through these two approaches, we can confirm that considering antenna pattern and satellite pass geometry, analytical models can be built to predict the time and duration of interference based on the angular separation between the two satellites (Cakaj et al. 2010a).

If operational impacts become severe, efforts must be performed to mitigate interference, specifically under low elevation. Figure 9.20 shows the case at elevation of 9°. Modulation index and root mean square index variations during the whole satellite pass, manifested in Doppler shift curve, confirm interference during the whole pass under the maximal elevation (Max-El) of 9°, which will make it much more difficult to determine a distress location, under these circumstances given in Figure 9.20.

Canadian team within a COSPAS-SARSAT system developed a procedure to interrupt RF transmission from the satellite with a minimal chance of irrecoverable failure. The NOAA-SARSAT executed this procedure when the operational impacts of interference became evident. The NOAA-SARSAT analyzed the downlink characteristics during the periods both before and after the mitigation actions were taken. This process is further described.

Satellite pair S10/S12, as the worst case of adjacent interference is further analyzed. The Canadian procedure to interrupt the downlink RF transmission from the satellite is considered to be applied as a method to mitigate adjacent satellite interference. The turnoff transmission was planned for S10. The situation before turning off the planned S10 satellite is given in Figure 9.21 under Max-El of 12°, where the interference presence shows up!

In Figure 9.21 before the turnoff of the S10 downlink, the received carrier frequency can be seen jumping from one satellite's downlink to the other one, causing the degradation of downlink capabilities. The modulation indices are higher during these times since the receiver cannot lock on only one carrier. Figure 9.22 shows the same plot after the downlink of S10 had been turned off. They show that the only increase of the modulation indices occurs near the LOS, when the signal is

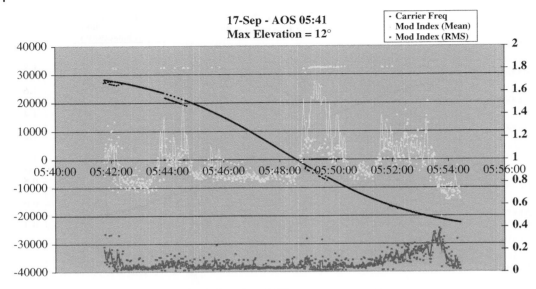

Figure 9.21 Doppler curve for Maximal elevation of 12°.

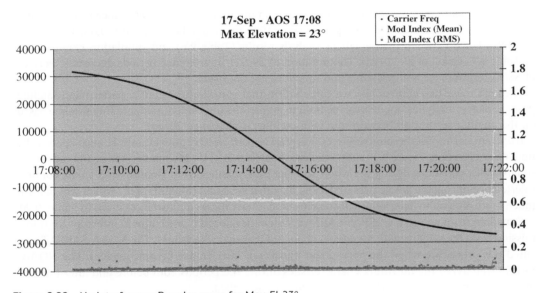

Figure 9.22 No interference Doppler curve for Max-El 23°.

the weakest. Figure 9.22 is typical of what would be seen during a nominal pass with no interference. The procedures developed by Canada and executed by the USA were successful in interference mitigation (Cakaj et al. 2010a).

Conclusion: It is confirmed that adjacent SARSAT satellites with short differences in orbital period interfere with each other. During these interference periods, significant degradation of downlink occurs. The procedure to interrupt the downlink RF transmission from the "undesired" satellite is applied as a method to mitigate adjacent satellite interference. For newly built terminals though, larger antennas with a narrower beamwidth may also reduce the adjacent interference issue and impacts.

The DASS (Distress Alert Satellite System) is a newly developed & future approach intended to enhance the international COSPAS-SARSAT program. In this effort the satellite-aided search and rescue (SAR) system will install 406 MHz SAR instruments on the MEO navigational satellites [GPS (US), Galileo (EU), and Glonass (Russian Federation)]. With an expected 80 satellites once fully operational, new processing algorithms and interference mitigation strategies should also be considered. Because of the much higher altitudes of MEO satellites, a larger separation distance exists, and the adjacent interference will be less pronounced.

9.8 Uplink Interference Identification for LEO Search and Rescue Satellites

For a rescue operation to be successful, it is crucial that the distress location be rapidly determined. The location determination is based on Doppler frequency shift, and its accuracy depends on the signal quality received at the satellite from the distress beacon. The distress beacon signal can be disturbed by interferers, which degrade the performance of the on-board 406 MHz Search and Rescue Processor (SARP) and reduce the probability of detecting real beacon messages (COSPAS – SARSAT System Monitoring and Reporting 2008). Emergency distress beacons are essentially specialized radio transmitters for search and rescue purposes carried by airplanes, ships, and individuals. SAR (search and rescue) satellites can even "hear" faint distress signals from beacons. The beacon can be activated manually or automatically. Since February 2009, rescue beacons have transmitted on 406 MHz. Some characteristics of 406 MHz beacons are shown in Table 9.8 (Specification for COSPAS – SARSAT406MHz Distress Beacons 2008).

The 406.0–406.1 MHz band has been allocated by the International Telecommunication Union (ITU) for distress alerting using low-power, emergency radio beacons. This 100 KHz licensed band is organized in channels for multiuser random access at the satellite. Frequency spectrum lies in the band 406.0–406.1 MHz. These radio beacons emit stable, constant frequency, which is highly important for location determination. The 406 MHz carrier is phase modulated with information such as beacon identification, synchronization frame, and the nature of emergency. From the time slot of 500 ms, only 160 ms are dedicated for poor carrier and the rest is for modulated data, such are beacon type, its country of origin and the registration number of the maritime vessel, aircraft, or individual (Specification for COSPAS – SARSAT406MHz Distress Beacons 2008).

Idea: If the beacon signal at the distress location to be sent through the uplink to the satellite is disturbed, the satellite's receiver will not be able to receive the appropriate accurate (clean) beacon signal, and consequently will not yield the right Doppler curve to provide the correct rescue location. The question is to identify the impact of uplink interference on Doppler curve, and consequently on distress location determination (Cakaj et al. 2010b).

Table 9.8 Some of beacon characteristics.

Output power	5 W
Transmission	Burst (500 ms on every 50 sec)
Modulation	Phase
Frequency stability	High

Method: For confirmation of the above discussed effect, the records are taken from NSOF (NOAA Satellite Operation Facility), where the author did his research (NSOF 2010; National Telecommunications and Information Systems 2006).

Active signal sources in various areas of the world can produce spurious emissions in the 406 MHz allocated range, interfering with the NOAA-SARSAT system, whether or not they operate at the 406 MHz band. How severe the interference is will depend on the amount of frequency overlap between the interfering spectrums and the allocated channel passband, relative to the tolerable level of the receiver. NOAA-SARSAT satellites have 406 MHz repeaters for retransmitting emissions received from Earth in the band 406.0–406.1 MHz. As a result, the time/frequency pairs of interference emissions can be measured with specially designed LEOLUTs, which monitor 406 MHz interference using specific software in the LEOLUTs in conjunction with a 406 MHZ SAR processor. The software at the LEOLUT utilizes Fast Fourier Transform (FFT) analysis to determine time, frequency, and the power of the interfering signal.

Results: Two spectrograms are further given, extracted by the examination of the Doppler frequency shift of data transmissions from detected beacons or interfering transmitters (NSOF 2010; USTTI, Course M6-102, National Telecommunications and Information Systems 2006; Cakaj et al. 2010b).

Figure 9.23 shows the spectrogram view with no interference. The Doppler curve is related to beacon A. The number of data points (Doppler events) depends on communication duration between a satellite and LEOLUT. Satellite orbital parameters, the duration, and, consequently, the expected data points are known in advance. In Figure 9.23, Doppler points are discrete, since the beacon hits the satellite in intervals of 50s. From Figure 9.23, it is obvious a good Doppler curve was provided during the satellite pass and the distress location was obtained.

Figure 9.24 shows the spectrogram view when interference is present. Two beacons, B and C, are disturbed by interference. For beacon B, nine data points were expected and for beacon C, seven data points were expected. This indicates a medium duration of communication between a satellite

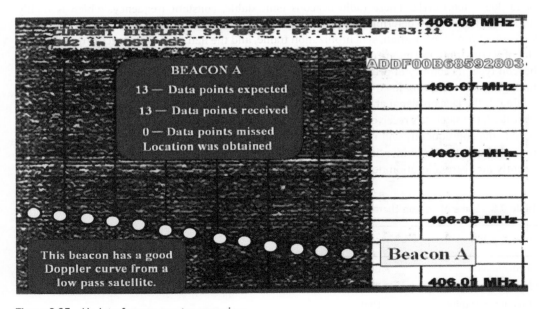

Figure 9.23 No interference spectrogram view.

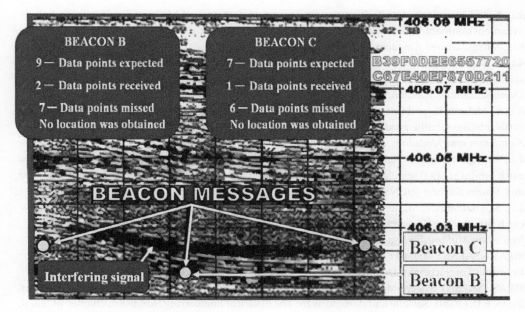

Figure 9.24 Interference spectrogram view.

and a LEOLUT. The interfering signal is shown in Figure 9.24. The interfering signal is continuous for a long period of time as compared to the periodic 500 ms beacon bursts. The continuous interfering signals may produce a Doppler curve, but no identification code can be extracted from an interfering signal since its modulation, if any, would not be in the correct format. The lack of identification code is an interference indicator. Interference caused insufficient data points to be captured for a Doppler curve, and, consequently, no distress location for these two beacons was obtained.

Conclusion: NOAA-SARSAT is a data communication system dedicated for search and rescue purposes, oriented on determination of distress locations worldwide. Since uplink interference can disturb the satellite's receiver, blocking the location of the distress location, thus a very sophisticated method has been developed and applied at LEOLUTs to identify uplink interference, providing spectrograms above applied. This method provides data about frequency, power, and time of interference.

References

Bulloch, C. (1987) *Search and Rescue by satellite – Slow steps toward an operational system*, (ISSN 0020-5168), 42: 275 -277.

Cakaj, S. (2010). Intermodulation interference modelling for low earth orbiting satellite ground stations book chapter. In: *Modelling, Simulation and Optimization* (ed. R.G. Rey and M.L. Muneta), 97–116. Croatia: INTECH.

Cakaj, S. (2012). Modulation index application for satellite adjacent downlink interference identification. In: *The 6th European Conference on Antennas and Propagation EUCAP 2012*, 2000–2004. Prague, Czech Republic: IEEE, March 26-30, 2012.

Cakaj, S. and Malaric, K. (2007). Rigorous analysis on performance of LEO satellite ground station in urban environment. *International Journal of Satellite Communications and Networking* 25 (6): 619–643, Surrey, United Kingdom.

Cakaj, S., Keim, W.A., and Malaric, K. (2005). Intermodulation by uplink signal at low earth orbiting satellite ground station. In: *18th International Conference on Applied Electromagnetics and Communications, ICECom*, 193–197. Dubrovnik, Croatia: IEEE.

Cakaj, S., Malaric, K., and Schotlz, L.A. (2008). Modelling of interference caused by uplink signal for low earth orbiting satellite ground stations. In: *17th IASTED International Conference on Applied Simulation and Modelling*, ASM 2008, June 23 –25, 187–119. Greece.

Cakaj, S., Fitzmaurice, M., Reich, J., and Foster, E. (2010a). The downlink adjacent interference for low earth orbiting (LEO) search and rescue satellites. *International Journal of Communications, Networks and System Sciences (IJCNS)* 3 (2): 107–115.

Cakaj, S., Fitzmaurice, M., Reich, J., and Foster, E. (2010b). Uplink interference identification for low earth orbiting (LEO) search and rescue satellites. In: *52nd International Symposium ELMAR 2010 focused on Multimedia Systems and Applications*, 173–176. Zadar, Croatia: IEEE.

COSPAS – SARSAT 406MHz Frequency Management Plan (2008) C/T T.012, Issue 1 – Revision 5, Probability of Successful Doppler Processing and LEOSAR System Capacity, October.

COSPAS –SARSAT System Monitoring and Reporting (2008) C/S A.003, Issue 1, Revision 15, October.

Difonzo, F.D. (2000). *Satellite and Aerospace*. In: *The Electrical Engineering Handbook*, chapter 74 (ed. R. C. Dorf). Boca Raton, FL: CRC Press LLC.

Dodel, H. (1999). *Satellitenkommunikation*. Berlin: Springer-Verlag.

Gordon, D.G. and Morgan, L.W. (1993). *Principles of Communication Satellites*. New York: Wiley.

Kanellopoulos, D.J., Kritikos, D.T., and Panagopoulos, D.A. (2007). Adjacent satellite interference effects on the outage performance of a dual polarized triple site diversity scheme. *IEEE Transaction on Antennas* 55 (7): 2043–2055.

Landis, S.J. and Mulldolland, J.E. (1993). Low-cost satellite ground control facility design. *IEEE, Aerospace & Electronics systems* 2 (6): 35–49.

Losik, L., (1995) Final report for a low–cost autonomous, unmanned ground station operations concept and network design for EUVE and other NASA Earth orbiting satellites, Technology Innovation Series, Publication 666, Center for EUVE Astrophysics, University California, Berkeley.

Maral, G. and Bousquet, M. (2002). *Satellite Communication Systems*. Chichester, England: Wiley.

Mendenhall, N. G. (2001) A Study of Intermodulation between Transmitters Sharing Filterplexed or Co-Located Antenna Systems. Engineering Broadcast Electronics, Inc., Quincy, IL.

NSOF (2010), https://commerce.maryland.gov/Documents/BusinessResource/NSOF-NOAA-Satellite-Operations-Facility.pdf

National Telecommunications and Information Administration, (2006) USTTI Radio Spectrum Frequency Management Course M6-102, 2006, Washington, DC.

Richharia, M. (1999). *Satellite communications systems*. New York: McGraw Hill.

Root-Mean-Square 2012, http://mathworld.wolfram.com/Root-Mean-Square.htm

Sklar, B. (2001). *Digital Communication*. Englewood Cliffs, NJ: Prentice Hall PTR.

Specification for COSPAS – SARSAT406MHz Distress Beacons (2008) C/T T.001, Issue 3 – Revision 9 October.

Vataralo, F., Emanuele, G., Caini, C., and Ferrarelli, C. (1995). Analysis OF LEO, MEO and GEO global Mobile satellite Systems in the Presence of interference and fading. *IEEE Journal on Selected Areas in Communications* 13 (2): 291–299.

Yafeng, Z., Zhigang, C., and Zhengxin, M. (2004). Modulation index estimation for CPFSK signals and its application to timing synchronization. In: *International Symposium on Multi-Dimensional Mobile Communications*, 2e, 874–877. Beijing, China.: IEEE.

10

Two More Challenges

10.1 Introduction to the Two Challenges

Finally, this book concludes with two challenges about which I'll be further engaged in my scientific study, aiming to solve and include them in the future academic and industrial literature. The first one is related to the free space loss compensation by receiver's dynamic bandwidth at the ground station, which is presented as a poster under the title "Downlink Free Space Loss Compensation through Receiver Bandwidth Selectivity at LEO Satellite Ground Stations" in *IEEE SCVT 2016, 23rd Symposium on Communications and Vehicular Technology in the Benelux*, Mons in Belgium, 2016. The next one is related to the implementation of new parameters for better description of LEO satellite-ground station geometry, which is published under the title "The Parameters Comparison of the *Stralink* LEO Satellites Constellation for Different Orbital Shells," in *Frontiers in Communications and Networks-Aerial and Space Networks*, Vol.2, Article 643095, 2021.

10.2 Downlink Free Space Loss Compensation

The function of the satellite ground station is to receive or transmit the information from/to the satellite in the most reliable manner while retaining the desired signal quality. The demands for satellite services are increasing every day, especially in the field of providing broadband services (Botta and Pescape 2013). Early designs were focused on making the system suitable for commercial operations, but new generations have to develop ways of maximizing the downlink data throughput related to the broadband service requirements without significantly affecting the mission cost. Therefore, future satellite payloads will have to become more flexible and shall provide capacity at the lowest cost. Tunable filters could be a relevant option to simplify payload architectures by providing more flexibility on the frequency plan and/or bandwidth (Acquaroli and Morelli 2005). Nowadays, specifically for small and lightweight satellites are applied agile payload architectures with variable satellite effectively isotropic radiated power (EIRP) and reconfigurable coverage in order to maximize satellite mission performance (Leblond et al. 2009).

The major loss in communication between the low Earth orbit (LEO) satellite and the ground station is the free space loss. Free space loss varies since the distance from the ground station to the LEO satellite varies over time. Free space loss variation is usually compensated through variable satellite transmit power (EIRP) toward downlink. Intending to avoid the complexity due to the design and construction of the satellite, the challenging question is, can the free space loss be compensated through the receiver tunable bandwidth at the ground station?

Ground Station Design and Analysis for LEO Satellites: Analytical, Experimental and Simulation Approach, First Edition. Shkelzen Cakaj.
© 2023 The Institute of Electrical and Electronics Engineers, Inc. Published 2023 by John Wiley & Sons, Inc.

Idea: To mathematically prove that the free space loss can be compensated through receiver's bandwidth dynamic adjustability (tunability) at the ground station! Technical implementation to be further analyzed in the future!

Method: Math analysis and simulation approach are applied, considering the altitudes from 600 to 1200 km (Kamo et al. 2016). For downlink budget calculations, of the greatest interest is the receiving system signal-to-noise ratio [(S/N) or (S/N_0)] expressed by range equation (Sklar 2005), as:

$$\frac{S}{N_0} = \frac{EIRP(G/T_S)}{kL_SL_0} \tag{10.1}$$

where EIRP is effectively isotropic radiated power from the satellite transmitter. Considering that $N = N_0 \cdot B$, $N_0 = kT$ where, N_0 is spectral noise density, B ground station receiver bandwidth, $k = 1.38 \cdot 10^{-23}$ W/HzK is Boltzmann's constant, expressed in (dB) yields:

$$\frac{S}{N_0}(dB/Hz) = EIRP - L_S - L_0 + G/T_S + 228.6 \tag{10.2}$$

$$(S/N) = (S/N_0) - B \tag{10.3}$$

L_s is free space loss, L_0 denotes other losses (atmospheric loss, polarization loss, misspointing, etc.) and G/T_S is Figure of Merit. The reception quality of the satellite receiving system is commonly defined through a receiving system Figure of Merit as G/T_S (Sklar 2005; Cakaj and Malaric 2006):

$$T_S = T_A + T_{\text{comp}} \tag{10.4}$$

where G is receiving antenna gain, T_S is receiving system noise temperature, T_A is antenna noise temperature, and T_{comp} is composite noise temperature of the receiving system, including lines and equipment (Cakaj and Malaric 2006). The Figure of Merit G/T_S expresses the impact of external and internal noise factors.

Free space loss (L_S) is the greatest loss in transmitted power due to the long distance between the satellite and a ground station (see Eq. (1.13)). This loss is displayed as:

$$L_S = \frac{(4\pi d)^2}{\lambda^2} \tag{10.5}$$

where d is the distance (slant range) between the satellite and a ground station, and λ is the signal wavelength. The free space loss L_S is often convenient to be expressed as function of distance d and signal frequency f, and then L_S is

$$L_S(\varepsilon_0) = \left(\frac{4\pi f}{c}\right)^2 \cdot d^2(\varepsilon_0) \tag{10.6}$$

where $d(\varepsilon_0)$ is further represented by Eq. (10.7). Free space loss increases by both frequency and the distance.

$$d(\varepsilon_0) = R_E\left[\sqrt{\left(\frac{H + R_E}{R_E}\right)^2 - cos^2\varepsilon_0} - sin\,\varepsilon_0\right] \tag{10.7}$$

where H is the LEO satellite orbit altitude. Thus, the largest range is achieved under 0° elevation and the shortest range occurs at 90° elevation, since the satellite appears perpendicularly above the ground station. The range under the lowest elevation angle represents the worst link budget case, since that range represents the maximal possible distance between the ground station and the satellite. This range increases as satellite's altitude increases.

Table 10.1 LEO satellite ranges.

Orbital altitude [km]	H 600 [km]	H 800 [km]	H 1000 [km]	H 1200 [km]
Elevation (ε_0)	Range [km]	Range [km]	Range [km]	Range [km]
0°	2830	3289	3708	4088
10°	1942	2372	2770	3136
20°	1386	1765	2120	2453
30°	1070	1392	1701	1996
40°	886	1164	1436	1698
50°	758	1005	1248	1486
60°	697	905	1129	1348
70°	680	847	1058	1266
80°	626	809	1012	1214
90°	600	800	1000	1200

Further, the free space loss is simulated considering altitudes of 600, 800, 1000, and 1200 km as typical LEO altitudes. For these altitudes applying Eq. (10.7) it is calculated the range from a hypothetical ground station to the satellite at appropriate altitudes, and presented at Table 10.1 (Cakaj et al. 2011). Table 10.1, shows that the distance (slant range) between the LEO satellite and the ground station changes and depends on elevation angle. This causes different free space loss and consequently different S/N ratio and the downlink margin (DM), respectively.

The downlink margin (*DM*) is defined as:

$$DM = (S/N)_r - (S/N)_{rdq} \tag{10.8}$$

where the *r* indicates expected S/N ratio to be received at receiver, and *rqd* means required S/N ratio requested by customer, as a predefined performance criteria. The positive value of *DM* is an indication of a good system performance. In efforts to maintain a positive link margin, a trade-off among parameters of range equation is required (Sklar 2005).

Considering that the required signal-to-noise ratio is defined as a fixed value, the idea behind this approach is to keep the downlink margin constant over different distances between the LEO satellite and the appropriate ground station. This means, keeping received *S/N* ratio always constant. The free space loss caused because changes on distance, it is intended to be compensated through receiver bandwidth tunability, respectively, with receiver selectivity.

The described case, mathematically expressed in decibels [dB] follows:

$$\left(\frac{S}{N}\right)_{\varepsilon_0=0} = \left(\frac{S}{N}\right)_{\varepsilon_0=x} = \left(\frac{S}{N}\right)_{\varepsilon_0=x+a} = \left(\frac{S}{N}\right)_{\varepsilon_0=90} = const. \tag{10.9}$$

where x and $(x + a)$ represent two neighbor temporary elevation (ε_0) values in the range from 0° to 90°, and a is defined as an incremental elevation step.

Applying Eq. (10.3) at Eq. (10.9), for the most general case, yields out:

$$\left(\frac{S}{N_0}\right)_{\varepsilon_0=x} - B_x = \left(\frac{S}{N_0}\right)_{\varepsilon_0=x+a} - B_{x+a} \tag{10.10}$$

where B_x, B_{x+a} refer to the receiving reconfigurable bandwidth when the LEO satellite is seen from the ground station under elevation of $x°$ and $(x+a)°$, respectively.

Considering that the transmit power (EIRP) from the LEO satellite toward the ground station is unchangeable under different elevations, other losses are also the same, and the ground station performance expressed through figure of merit is kept constant all the time. Then applying Eq. (10.2) for elevation of $x°$ and $(x+a)°$ the ratio signal over noise density (range equation) expressed in decibels [dB] is:

$$\left(\frac{S}{N_0}\right)_{\varepsilon_0 = x} = EIRP - L_{S(\varepsilon_0 = x)} - L_0 + G/T_S + 228.6 \tag{10.11}$$

$$\left(\frac{S}{N_0}\right)_{\varepsilon_0 = x + a} = EIRP - L_{S(\varepsilon_0 = x + a)} - L_0 + G/T_S + 228.6 \tag{10.12}$$

where $L_{S(\varepsilon_0 = x)}$ and $L_{S(\varepsilon_0 = x + a)}$ refers to the free space loss when the LEO satellite is seen from the ground station under elevation of $x°$ and $(x+a)°$. Applying Eqs. (10.11) and (10.12) at Eq. (10.2) yields (Kamo et al. 2016):

$$L_{S(\varepsilon_0 = x + a)} - L_{S(\varepsilon_0 = x)} = B_x - B_{x+a} \tag{10.13}$$

Since $L_{S(\varepsilon_0 = x)} \triangleright L_{S(\varepsilon_0 + a)}$ (See Table 10.1), ΔL_{Sx} is denoted as:

$$\Delta L_{Sx} = L_{S(\varepsilon_0 = x)} - L_{S(\varepsilon_0 + a)} \tag{10.14}$$

and leads to the equation for the free space loss compensation through the receiving bandwidth adjustability as:

$$B_{x + a} = B_x + \Delta L_{Sx} \tag{10.15}$$

Applying Eq. (10.6) at Eqs. (10.14) and (10.15) yields out:

$$L_{S(\varepsilon_0 = x)} - L_{S(\varepsilon_0 = x + a)} = \Delta L_{Sx} = 20\,log\left(\frac{d(\varepsilon_0 = x)}{d(\varepsilon_0 = x + a)}\right) \tag{10.16}$$

$$B_{x + a} = B_x + 20\,log\left(\frac{d(\varepsilon_0 = x)}{d(\varepsilon_0 = x + a)}\right)\ (dBHz) \tag{10.17}$$

where $d(\varepsilon_0 = x)$ and $d(\varepsilon_0 = x + a)$ refers to the distance (slant range) from the ground station to the satellite when the satellite is seen from the ground station under elevation of $x°$ and $(x + a)°$ (Kamo et al. 2016).

The LEO satellite path over the ground station is characterized with two typical events such as the satellite acquisition and the satellite loss. Both of these events happen under elevation of $0°$ having the longest distance between the ground station and the satellite. Thus, for these events, considering Eq. (10.7), the distance between the ground station and the satellite is

$$d(\varepsilon_0 = 0) = \sqrt{H(H + 2R_E)} \tag{10.18}$$

representing the longest distance, which causes the highest free space loss and consequently decreasing ratio signal over noise density (S/N_0). Under these circumstances, the receiving bandwidth is denoted as B_0, where index zero is referred to elevation of $0°$.

Another typical event of the LEO satellite path is when the satellite is seen perpendicularly above the ground station. This happens under elevation of $90°$, having the shortest possible distance between the satellite and the ground station. Applying Eq. (10.7) for the elevation of 90° yields:

$$d(\varepsilon_0 = 90) = H \tag{10.19}$$

This represents the shortest distance, which causes the lowest free space loss and consequently less impact on the signal-over-noise density ratio (S/N_0). Under these circumstances, the receiving bandwidth is denoted as B_{90}, where index ninety is referred to as elevation = 90°.

From above discussion stems that

$$\left(\frac{S}{N_0}\right)_{\varepsilon_0 = 0} \lhd \left(\frac{S}{N}\right)_{\varepsilon_0 = 90} \tag{10.20}$$

and consequently, in order to keep always constant S/N, it must be:

$$B_0 \lhd B_{90} \tag{10.21}$$

Further considering $x = 0°$ as an initial elevation and $a = 90°$ as an elevation step and then applying at Eq. (10.6), we get:

$$B_{90} = B_0 + 10 \log \left(1 + \frac{2R_E}{H}\right) \tag{10.22}$$

Since value under log is always positive, Eq. (10.22) confirms the conclusion at Eq. (10.21).

Considering all above elaboration, in order to keep S/N constant all over the LEO satellite orbit path, the receiver bandwidth should compensate the variation of free space loss. The farther the satellite is from the ground station (as lower elevation), the higher the free space loss. Thus, the receiving bandwidth must be narrowed to increase the selectivity level. Through bandwidth, this is expressed as:

$$B_0 \lhd B_x \lhd B_{x+a} \lhd B_{90} \tag{10.23}$$

So, the bandwidth must be narrowed (enhancing the selectivity) as the elevation decreases, in order to keep S/N constant over time and consequently the downlink margin (DM). Thus, downlink margin could be kept constant with receiver tunable bandwidth at the ground station.

Results: For different LEO orbit altitudes H, the difference between the widest (B_{90}) and the narrowest (B_0) applied bandwidth is presented in Table 10.2 (Kamo et al. 2016).

Conclusion: For typical LEO altitudes on range of 600–1200 km, for free space loss compensation, it is confirmed that the bandwidth should be tuned from B_0 up to $B_0 + 13.5\, dB$ and $B_0 + 10.6\, dB$, respectively, keeping constant downlink margin. The bandwidth must be narrowed (enhancing the selectivity) as the elevation decreases, in order to keep constant over time S/N and consequently the downlink margin.

Table 10.2 The difference between the widest and the narrowest applied bandwidth.

Orbital altitude [km]	$B_{90} - B_0 = 10 \log \left(1 + \dfrac{2R_E}{H}\right)$
$H = 600$ km	13.5 dB
$H = 800$ km	12.3 dB
$H = 1000$ km	11.4 dB
$H = 1200$ km	10.6 dB

Further research should focus on the correlation between bandwidth tuning and capacity. This section presents only the mathematical model, which requires signal processing related to continuous bandwidth change, and that, in turn, requires additional intelligence at the receiver for such configuration as the first challenge for the further study! Correlation between variable bandwidth and appropriate capacity remains a crucial research point. Trade-offs are mandatory!

10.3 Horizon Plane Width: New Parameter for LEO Satellite Ground Station Geometry

Different orbits are applied for different purposes. They are defined by their respective parameters, which determine their space geometry and on-ground coverage (footprint). Since LEOs, as the subject of this book are circular, for further discussion only circular orbits are considered. Among other parameters which describe the circular orbit, is included the ratio $(H + R_E)/R_E$, where H represents the satellite's altitude and R_E is the Earth's radius.

Table 10.3 presents the comparison of geometry, coverage, and some relevant parameters for typical LEO, middle Earth orbit (MEO), and geostationary Earth orbit (GEO). Data on Table 10.3 are calculated under these considerations: systems have been seen under elevation angle $\varepsilon_o = 10°$, then Earth's radius $R_E = 6378$ km, Earth's surface area $a_e = 511.2 \cdot 10^6$ km^2, and eccentricity $e = 0$ (Difonzo 2000).

All parameters of Table 10.3 have been defined and elaborated in previous chapters. Here we are concerned with $(H + R_E)/R_E$, which is used to determine the ratio between hypothetical spatial sphere of radius $(H + R_E)$ to the Earth's spherical body of radius R_E, all of these to create a view or feeling about dimensions in space for different orbits. Comparing values of $(H + R_E)/R_E$, from Table 10.3 (non-unit values: 1.12; 2.63; and 6.61, respectively, for LEO, MEO, and GEO), they really do provide sufficient difference to create a view about their comparative real dimension in space.

Table 10.3 Comparison of parameters for LEO, MEO, and GEO orbits.

Orbit	LEO	MEO	GEO
System	Iridium	ICO	INTELSAT
Inclination i (°)	86.4	45	0
Altitude H (km)	780	10 400	35 786
Semi major axis a (km)	7 159	16 778	42 164
Orbit period (min)	100.5	360.5	1436.1
$(H + R_E)/R_E$	1.12	2.63	6.61
Earth central angle β_0 (°)	18.6	58.0	71.4
Nadir angle α_0 (°)	61.3	22	8.6
Slant range d (km)	2325	14 450	40 586
One way time delay (ms)	2.6	51.8	139.1
Fraction of covered Earth's area	0.026	0.235	0.34

Table 10.4   Parameter $(H + R_E)/R_E$ for different shells.

Orbital altitude	340 km	550 km	1110 km
$(H + R_E)/R_E$	1.053	1.086	1.174

Idea: Satellite services are increasing every day, mostly provided by LEO satellites. The idea is to analyze whether the parameter $(H + R_E)/R_E$, provides adequate difference to reflect the difference about the orbits dimension in space, for different LEO altitudes. The intention is to confirm that this parameter makes sense if the altitudes differ seriously; otherwise, this parameter does not provide sufficient inputs if altitudes are too close to each other. Thus, the idea is to implement another parameter that, on the first reading, will provide a better sense for the orbital position-dimension in space for too-close orbital altitudes. This parameter is defined as ideal horizon plane width (*IHPW*), and its derivative is designed horizon plane width (*DHPW*), further elaborated here.

Method: Mathematical confirmation. For further analysis we apply three orbital shells planned to be used by Starlink (SpaceX) constellation, as 340, 550, and 1110 km (Starlink 2020; Starlink Satellite Missions 2020). For these shells, the parameter $(H + R_E)/R_E$ is given in Table 10.4.

The values 1.053; 1.086; and 1.174, respectively, for shells of 340, 550, and 1110 km do not provide sufficient information to draw conclusions about the space relation for the above considered shells. Further it is approached different view of interpretation, applying horizon plane width for above considered shells.

Another approach is based on the concept of the ideal and designed horizon plane. The events AOS (acquisition of the satellite) and LOS (loss of the satellite), at elevation of 0°, geographically determine the *ideal horizon plane*. The first event identifies the case when the satellite appears just at the horizon plane to be locked and communicate with the ground station (user) and the second one the case when the satellite just disappears from the horizon plane, being unlocked and no more in communication with the ground station (user). The virtual line connecting points in space when AOS and LOS happen at 0° elevation determines the *IHPW*. The width of the horizon plane depends on satellite's orbital altitude (Cakaj 2021). The horizon plane width in fact determines the ideal zone (area) where the ground station and the satellite can communicate.

The ideal horizon plane in fact represents the visibility region under 0° of elevation angle. But, because of different barriers (natural or artificial), this visibility region is often hidden, and consequently no communication can be locked/unlocked in between the satellite and the user at that elevation. Thus, the designers must predetermine the lowest possible elevation of the horizon plane for the safe communication to be locked/unlocked.

The horizon plane with appropriate *designed elevation* ($X°$) is considered the designed horizon plane, with its *DHPW*. Normally, the designed *DHPW* is shorter than the *IHPW* (Cakaj 2021). *IHPW* and *DHPW* are given in Figure 4.17.

These parameters sound good for expressing the LEO satellites geometry – more accurately, the zone (area) of communication between the satellite and the ground station. These parameters and their mathematical and geometrical correlation were analyzed in detail under Section 4.6, where we also calculate the *IHDW* and *DHPW* under 4.6 for three orbital shells (see Table 10.5), as well as $(H + R_E)/R_E$.

The ideal horizon plane is an ideal characteristic of any ground station under the LEO satellite coverage area. The designed horizon plane is determined by designed elevation angle (known as the

Table 10.5 Parameters $(H + R_E)/R_E$, *IHPW*, and *DHPW* for different shells.

Orbital altitude	340 km	550 km	1110 km
$(H + R_E)/R_E$	1.053	1.086	1.174
IHPW ($\varepsilon_0 = 0°$) [km]	4206.8	5405.8	7833.9
DHPW ($\varepsilon_0 = 40°$) [km]	775.9	1240.1	2405.1

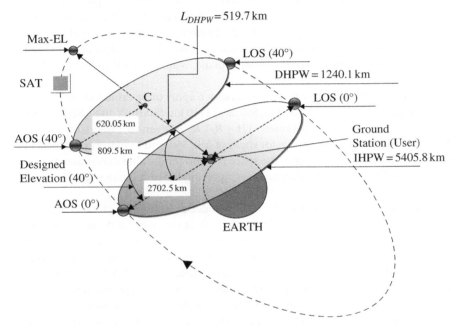

Figure 10.1 *IHDW, DHPW, d_{max},* and *L_{DHPW}* dimensions for *H* = 550 km.

masking elevation angle), identifying the communication zone between a LEO satellite and the ground station. Thus, two more involved parameters in Table 10.5 seem to provide better understanding about the geometry in space of an LEO satellite flying under altitude *H*, above the ground station within its footprint. An LEO satellite at altitude *H* = 550 km and elevations of 0° and 40° as seen from the ground station are presented in Figure 10.1 (Cakaj 2021).

For better understanding of the LEO satellite geometry, in the future, it is likely that parameters *IHPW* and *DHPW* will be a part of comparison tables for LEO satellites and appropriate ground stations, including academia and industry literature. This is given in Table 10.6, where two shaded lines present two new geometrical parameters added to Table 10.3 (Cakaj 2021).

Conclusion: Implementation of *IHPW* and *DHPW* parameters will serve as new tools for better understanding of LEO satellites and the appropriate ground stations geometry. They will provide better sense of dimension of the communication area between LEO satellite and the ground station. These parameters support better spatial view.

Table 10.6 Comparison of parameters for LEOs under different shells.

Parameters	The first shell H = 550 km	The second shell H = 1110 km	The third shell H = 340 km
Radius r [km]	6921	7481	6711
Velocity v [km/s]	7.589	7.299	7.706
Orbital period T [min]	95.5	107.3	91.2
Number of daily passes n	15.03	13.38	15.74
$(H + R_E)/R_E$	1.086	1.174	1.053
Slant range d_{max} [km]	809.5	1569.9	506.5
One way time delay τ [ms]	2.69	5.23	1.68
IHPW [km]	5405.8	7833.9	4206.8
DHPW [km]	1240.1	2405.1	775.9
Nadir angle α_0 (°)	44.8	40.7	46.6
Earth's central angle β_0(°)	5.2	9.3	3.4
Fraction of Earth's area [%]	0.206	0.657	0.088

References

Acquaroli, L. and Morelli, B. (2005) *Agile Payload Development and Testing of the Simplified Qualification Model*. The European Conference on Spacecraft Structures, Materials and Mechanical Testing, Noordwijk, The Netherlands.

Botta, A. and Pescape, A. (2013) New Generation Satellite Broadband Internet Service: Should ADSL and 3G Worry. *TMA 2013,* co-lacted with IEEE INFOCOM 2013. Turin, Italy.

Cakaj, S. (2021). The parameters comparison of the "Starlink" LEO satellites constellation for different orbital shells. *Frontiers in Communications and Networks-Aerial and Space Networks* 2: 643095.

Cakaj, S. and Malaric, K. (2006). Composite noise temperature at low earth orbiting satellite ground station. In: *International Conference on Software, Telecommunications and Computer Networks, SoftCOM 2006*, 214–217. Split, Croatia: IEEE.

Cakaj, S., Kamo, B., Kolici, V., and Shurdi, O. (2011). The range and horizon plane simulation for ground stations of low Earth orbiting (LEO) satellites. *International Journal of Communications, Networks and System Sciences (IJCNS)* 4 (9) (September 2011): 585–589.

Difonzo, D.F. (2000). Satellite and aerospace. In: *The Electrical Engineering Handbook*, Chapter 74 (ed. R. C. Dorf), 1857–1873. Boca Raton: CRC Press LLC.

Kamo, B., Cakaj, S., and Agastra, E. (2016). Downlink free space loss compensation through receiver bandwidth selectivity at LEO satellite ground stations. In: *Poster, IEEE SCVT 2016, 23rd Symposium on Communications and Vehicular Technology in the Benelux*, 1–5. Mons, Belgium.

Leblond, H. et al. (2009). *When New Needs for Satellite Payloads Meet with New Filters Architecture and Technologies*. European Microwave Integrated Circuits Conference, Rome, Italy, pp. 359–362.

Sklar, B. (2005). *Digital Communication*. New Jersey: Prentice Hall PTR.

Starlink (2020). https://en.wikipedia.org/wiki/Starlink#Global_broadband_Internet (accessed 17 January 2021).

Starlink Satellite Missions (2020). https://directory.eoportal.org/web/eoportal satellite-missions/s/ starlink (accessed 17 January 2021).

11

Closing Remarks

Two scientific-industrial achievements of the last century, satellite systems and broadband internet, enabling multimedia communications and impacting society not only on communication, but also on education, industry, economy and all real-time aspects of the socio-economic development, moved the world toward globalization. The internet has become an indispensable factor for the social and economic transformation in the future.

I attended the World Broadband Forum on 2014 in Amsterdam, and I was positively surprised with initiative that internet access should be considered among "elementary human rights" applicable by convents of UN (United Nations). Unfortunately, based on statistics of early 2021, around 47% of world population was still without internet services (The state and the future of LEO satellite Internet Connectivity in Africa, 2022). But worldwide efforts are progressing toward worldwide broadband internet services.

Communications-integrated satellite-terrestrial networks used for global broadband services have gained a high degree of interest from scientists and industries worldwide, for their potential to provide ubiquitous coverage to broadband internet access services. The most convenient structures for such use are low Earth orbits (LEO) satellites, since they fly closer to the Earth compared to the other orbits, and consequently provide significantly lower latency, which is essential for reliable and safe communications.

Active satellite projects related to an integrated satellite-terrestrial communications network are Iridium constellation with 66 satellites (Cochetti, 2015), OneWeb constellation with 648 satellites (De Selding, 2015), (Pultarova and Henry, 2017). Amazon has received approval from the Federal Communications Commission (FCC) to launch 3236 spacecraft in its Kuiper constellation (Howell, 2020) and Telesat has been approved for a network of 117 spacecrafts (Foust, 2018), but recently the most serious activities have been undertaken by SpaceX company.

The Starlink satellites constellation has been developed and partly deployed by the US company SpaceX. The constellation is planned to be organized in three spatial shells, each made up of several hundred small-dimensioned and lightweight LEO satellites specially designed to provide broadband coverage across the entire Earth through their interoperability, combined with the ground stations, as a part of the satellite-terrestrial integrated network (Starlink Satellite Missions, 2020).

By April 1, 2020, the Union of Concerned Scientists (UCS), which maintains a database of active satellites in orbits, declared a total of 2,666 satellites in space, with 1918 in low Earth orbits (Geospatial World, 2021). Through the efforts of SpaceX and other companies, in the near future, satellites will envelop the Earth. This will surely impact the sky and will be challenging for the future of scientific research.

Ground Station Design and Analysis for LEO Satellites: Analytical, Experimental and Simulation Approach, First Edition. Shkelzen Cakaj.
© 2023 The Institute of Electrical and Electronics Engineers, Inc. Published 2023 by John Wiley & Sons, Inc.

I have been engaged in telecommunications for more than 40 years, including industry and academia. When I began my career, communicating involved writing letters and dialing rotary telephones. In the early 1990s, the internet transformed technology, and for about the last 20 years LEO micro-satellites have taken communications forward again, as satellites envelope the Earth with the potential to provide broadband services worldwide.

This represents a gigantic technological step for worldwide equality, but brings with it a lot of challenges in the future. Satellites will simplify communications for every point on Earth, including yards, schools, factories, trains, bus stations, golf sites, gas stations – simply everywhere! Let me close the book with two examples.

November 17, 2020, *Reddit* published the picture titled "Dish with Clear View of the Sky and Friendly Elk" (Reddit, 2020), given as Figure 11.1, which speaks for itself! The dish (ground station access point) and the elk coexist, not disturbing each other! Such satellites are making communications and internet access possible in the Ukraine.

Xin Yang submitted his PhD thesis in November 2018 titled "Low Earth Orbit (LEO) Mega Constellations-Satellite and Terrestrial Integrated Communication Networks" at the University of Surrey. Under chapter literature review, page 41, he has a picture captioned as "*Starlink* constellation (11,943 Satellites, Imaginary Representation, and Created with SaVi [55])" representing the enveloped Earth by Starlink satellites (Yang, 2018), shown in Figure 11.2 (Yang, 2018; Lutz et al., 2012).

The first picture speaks to the simplicity and the second one to the complexity of satellites. In between these two pictures, in between simplicity and complexity, there is much room for future scientific challenges – not only in terms of the potential for ubiquitous broadband services but also in terms of concerns regarding sky transformation.

Figure 11.1 Dish with clear view of the sky and friendly elk.

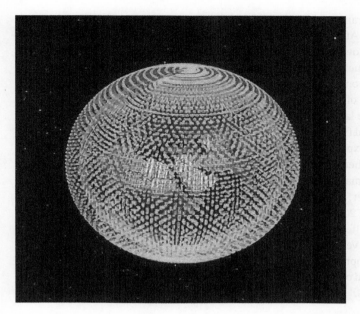

Figure 11.2 Starlink constellation (11 943 Satellites, Imaginary Representation, created with SaVi [55]) (Yang, 2018; Lutz et al., 2012).

Access points are not limited to the ground. IFBC (Inflight Broad Connectivity) is targeting Live-TV broadcasting and media streaming requiring ultra-high connectivity, in particular for intercontinental flights, as part of its 5G service. In this context, next-generation mobile networks (5G/6G) are setting design targets at 1.2Gpbs per aircraft. In January 2018, around 90 airlines had either installed or committed to install IFBC solutions. Offering connectivity was first seen as a differentiating factor; however, as more and more airlines provide connectivity, offering in-flight Wi-Fi starts to become a must-have in order to stay competitive in the extremely challenging airline market (Schneiderman, 2019). LEO satellites represent a good and convenient solution, being closer to airplane flight and providing a short signal delay. Iridium constellation has been widely adopted across all aviation segments to support communication on fast-moving aircraft. Broadband HTS (high-throughput satellite) is a means to provide both higher capacity and lower latency due to closer physical proximity of satellite to aircraft. In fact, the operator OneWeb is on track to be the world's first global LEO broadband systems provider to be operational by 2022 (Schneiderman, 2019). Providers are also implementing access points in vessels to marine satellite internet services while sailing the Seven Seas or moored at a dockside slip. The FCC has authorized SpaceX to provide user stations on boats, planes, and trucks (Sheetz 2022).

As the inflight and maritime broadband connectivity providers are offering Wi-Fi, the structure of the respective access points (virtual ground stations attached to vessels or airplanes) should follow the satellite's structure to be in full coordination with the appropriate LEO satellite network.

In closing, I would like to mention three important events that emphasize the satellite systems academic and industry achievements in the last 10 years.

The first one, April 26, 2014, the first stage of the Falcon 9 rocket made a controlled power landing on the surface of the Atlantic Ocean. It was returning from delivering an inflatable habitat into space for NASA. The inflatable room is attached to the International Space Station (ISS) for a two-year test and becomes the first such habitat for humans in orbit (Szondy, 2014).

Second, the Orbit Fab, a San Francisco-based space-industry startup, has developed end-to-end refueling service using its Rapidly Attachable Fluid Transfer Interface (RAFTI). RAFTI, Orbit Fab's first product, is a fueling port to allow satellites to be refueled easily in orbit. Orbit Fab's architecture includes a system of tankers and fuel tenders in low Earth orbit and geostationary orbit operational since 2018. The refuel of satellites on orbit is vital for operators since it allows them greater maneuverability and can extend the life of a mission (Lockheed Martin, 2022).

The third event is related to the launching of the satellites from the International Space Station (ISS). Typically, satellites launched from Earth require dedicated rockets to propel them into proper orbit. The Japan Aerospace and Exploration Agency found a way to cut the costs of this activity by designing a small satellite launcher, recently installed on the International Space Station (ISS) (Dempsey, 2000; Nogawa, and Imai, 2021). Commercial satellite deployment from the ISS has proven to be simpler, easier, faster, more secure, and cheaper. The last nanosatellite successfully launched from ISS on February 4, 2022, is UAE (United Arab Emirates) and Bahrain's Light-1 satellite, dedicated to studying the effects of gamma rays (Parks, 2022).

Finally, these activities confirm very serious engagement and investment of science and industry in satellites, followed by the appropriate access points or ground stations, thus, in this book, mathematical analysis, simulations, and on-site experiments were used to investigate the performance of access points and ground stations.

References

Cochetti, R. (2015). Low earth orbit (LEO) mobile satellite communications systems. In: *Mobile Satellite Communications Handbook*, 119–155. Hoboken, NJ: Wiley Telecom.

Dempsey, R. (2000) The International Space Station Operating an Outpost in the New Frontier, NASA. https://www.nasa.gov/sites/default/files/atoms/files/iss-operating_an_outpost-tagged.pdf

De Selding, B.P. (2015). *Virgin, Qualcomm Invest in OneWeb Satellite Internet Venture*. Paris: SpaceNews.

Foust, J. (2018) Data from Telesat to announce manufacturing plans for LEO constellation in coming months. *SpaceNews* (February 18). https://spacenews.com/telesat-to-announce-manufacturing-plans-for-leo-constellation-in-coming-months/

Geospatial World (2021) How many satellites orbit Earth? https://www. geospatialworld.net/blogs/how-many-satellites-orbit-earth-and-why-spacetraffic-management-is-crucial/ (Accessed January 17, 2021).

Howell, E. (2020) The FCC has approved Amazon's plan for its Kuiper satellite constellation. Here's what that means. Space.com (August 20), https://www.space.com/amazon-kuiper-satellite-constellation-fcc-approval.html.

Lockheed Martin (2022) Refueling satellites in space, https://www.lockheedmartin.com/en-us/news/features/2021/refueling-satellites-in-space.html#:~:text=Orbit%20Fab%2C%20a%20San%20Francisco,be%20refueled%20easily%20in%20orbit [Accessed March 2022].

Lutz, E., Werner, M., and Jahn, A. (2012). *Satellite Systems for Personal and Broadband Communications*. Berlin, Germany: Springer-Verlag.

Nogawa, Y. and Imai, S. (2021). *Cubesat Handbook: From Mission Design to Operations*, 445–454. Academic Press https://doi.org/10.1016/B978-0-12-817884-3.00024-2.

Pultarova, T. and Henry, C. (2017). *OneWeb Weighing 2,000 more Satellites*. Washington D.C.: SpaceNews.

Schneiderman, B. (2019) Innovations In-flight connectivity, Satellite Market & Research. http://satellitemarkets.com/innovations-flight-connectivity.

Starlink Satellite Missions (2020). https://directory.eoportal.org/web/eoportal/ satellite-missions/s/ starlink (Accessed January 17, 2021).

Sheetz, M. (2022). FCC authorizes SpaceX to provide mobile Starlink internet service to boats, planes and trucks. CNBC (June 30), https://www.cnbc.com/2022/06/30/fcc-approves-spacex-starlink-service-to-vehicles-boats-planes.html.

Szondy, D. (2014) SpaceX confirms successful Falcon 9 soft landing on the Atlantic Ocean. https:// newatlas.com/falcon-9-landing/31797/#:~:text=April%2026%2C%202014%20The%20first%20stage% 20of%20the,landing%20on%20the%20surface%20of%20the%20Atlantic%20Ocean [Accessed March 2022].

Parks, L. (2022) UAE and Bahrain's Light-1 nanosatellite successfully launches from ISS. https://www. itp.net/emergent-tech/uae-bahraini-nanosatellite-launches#:~:text=Emergent%20Tech%20February %204%2C%202022%20UAE%20and%20Bahrain%E2%80%99s,used%20to%20study%20the%20effects% 20of%20gamma%20rays[Accessed March 2022].

Reddit (2020) https://www.reddit.com/user/mscout206/ (Accessed November 15, 2020).

The state and the future of LEO satellite Internet Connectivity in Africa (2022), The special report by Space in Africa, commissioned by Via Satellite. http://interactive.satellitetoday.com/via/january-february-2022/the-state-and-future-of-leo-satellite-internet-connectivity-in-africa/[Accessed March 2022].

Yang, X. (2018) Low Earth orbit (LEO) Mega constellations – satellite and terrestrial integrated communication networks. Dissertation thesis. Institute for Communication Systems Faculty of Engineering and Physical Sciences University of Surrey Guildford, 196.

Index

Ground Station Design and Analysis for LEO Satellites: Analytical, Experimental and Simulation Approach,
First Edition. Shkelzen Cakaj.
© 2023 The Institute of Electrical and Electronics Engineers, Inc. Published 2023 by John Wiley & Sons, Inc.

Printed and bound by CPI Group (UK) Ltd, Croydon, CR0 4YY

16/04/2025

14658606-0004